CONTROL TECHNOLOGY

pupils' assignments

Control Technology was written by G.J. Fox and D.F. Marshall, Danum Grammar School for Boys, Doncaster, and was edited for Project Technology by G.L. Viles. It was revised P.W. Ghee, D. Hendley, A. Paul and G.L. Viles (editor).

CONTROL TECHNOLOGY

pupils' assignments

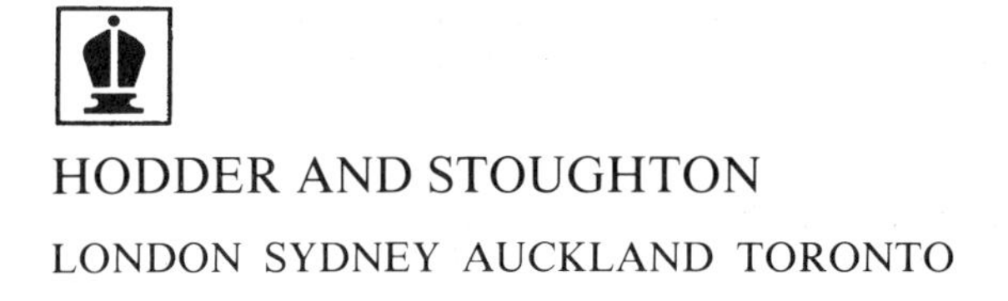
HODDER AND STOUGHTON
LONDON SYDNEY AUCKLAND TORONTO

British Library Cataloguing in Publication Data

Schools Council, *Project Technology*
Control technology pupils' assignments.——2nd ed.
1. Automatic control
I. Title II. Fox, G.J. III. Marshall, D.F.
629.8′0724 TJ213

ISBN 0 340 36406 8

First printed 1974,
Tenth Impression 1984
Second edition 1986
Second Impression 1987

Printed in Great Britain for Hodder and Stoughton Educational, a division of Hodder and Stoughton Ltd., Mill Road, Dunton Green, Sevenoaks, Kent, by Page Bros (Norwich) Ltd

Contents – pupils' assignments

Important note
In these assignments, you will see that some instructions have been printed in ***bold italic*** type. This indicates that you must write down in your notebook your observations or measurements from carrying out these instructions.

Foreword

Project Technology was a major curriculum development project initiated by the Schools Council to promote a better understanding by boys and girls in school of the importance and relevance of technology. The Project was concerned with helping teachers to stimulate an awareness of the material and scientific forces which effect change in our society and to develop knowledge of these forces and their means of control by the direct involvement of pupils in technological activities.

The Project Technology teaching-material programme is the result of a careful assessment of what is required in the schools, followed by trials and editing of the material itself. The Project Technology team felt that it was essential to draw on the experience, imagination and flair of individual teachers who, over a period of years, had developed technological work in particular parts of the school curriculum.

It is against this background that the *Control Technology* course hould be seen. Other teaching material, notably the Project Techiology Handbooks series, indicates both a thematic and a tactical ipproach to school technology. All of this reflects the diverse iature of the work being done and of the alternative teaching nethods and organisation which are possible.

This course, however, is intended to meet the needs of those chools who wish to develop a structured and sequential two- or hree-year course, covering an important area of technology. Some eachers have expressed the view that, while appreciating the value of what might be regarded as the more fortuitous involvement of pupils with technological projects and investigations, they should velcome a more systemic course approach.

The joint authors of *Control Technology*, Messrs G.J. Fox and).F. Marshall, have developed the course over a number of years at)anum Grammar School for Boys, Doncaster, where they were iven essential support and encouragement by the Headmaster (Mr :. Semper, OBE) and by the Doncaster Education Authority.

Clearly defined educational objectives were established, and the ppropriate teaching methods, based on pupil assignments, with ppropriate texts and equipment, were progressively developed, vith the support of the Project Technology team.

First-stage school trials were conducted in selected schools in the

Doncaster area with the generous help of the LEA. Secondary trials followed in various parts of the country. We are especially grateful to the teachers in these schools for the help and co-operation they provided for the authors, the Project Technology team (led by Mr G.L. Viles) and the Evaluators appointed by the Schools Council.

The *Control Technology* course has been carefully based on purpose-built equipment essential to the effective running of the course. A considerable effort has been put into the development of this equipment, at all stages, and we are indebted to all concerned, including the present suppliers. More recently, Trent Polytechnic are to be thanked for encouraging further development work by the National Centre for School Technology.

This revised edition has been prepared to serve the current examination syllabus set by the Associated Examining Board and the Joint Matriculation Board who at the time of writing, are the examiners for this subject at 16+.

Assignments using an electric motor have been amended for use with an improved model. Electronic assignments now follow current practise in circuit design, and the pneumatic component symbols now follow standards in current use.

Fluidic control is not included in current syllabuses and this original section has been removed. The logic section has been extended to meet current syllabus requirements and symbol notation.

Structures 1

1 With nuts and bolts, fasten any four pieces of Meccano strip together to form a quadrilateral similar to the one shown.

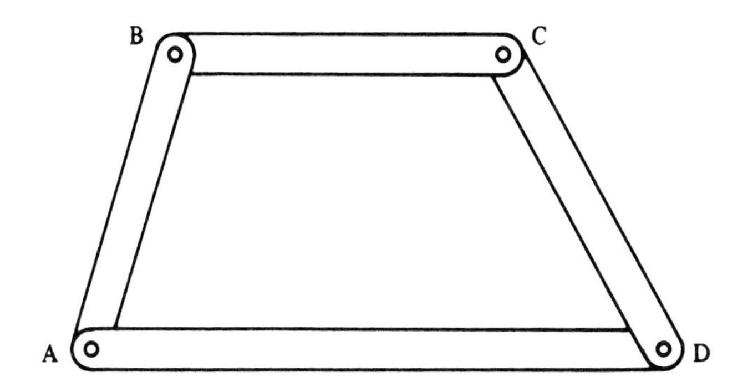

2 Try to push A and C towards each other.
Try to pull A and C apart.
Do the same with B and D.

In your notebook, draw line diagrams of your framework. Use arrows to indicate the direction of the forces and the position at which each is applied. Show with dotted lines the effect that the forces have on the shape of the structure.

3 ***Is the structure rigid?*** (i.e. will it retain its shape under the action of forces applied at any two points?)

4 If it is not rigid, how can it be made to retain a desired shape?

In your notebooks make sketches of as many alternative methods as possible.

5 When you have run out of ideas, analyse your sketches and select the solution which you think will solve the problem most satisfactorily.

Indicate which solution you have selected, giving reasons for your choice.

6 Build this solution into your model, re-test it for rigidity, and ***record in your notebooks the effectiveness of your modifications.***

Structures 2

1 Make a box-like structure of Meccano, using flat strips and angle brackets.

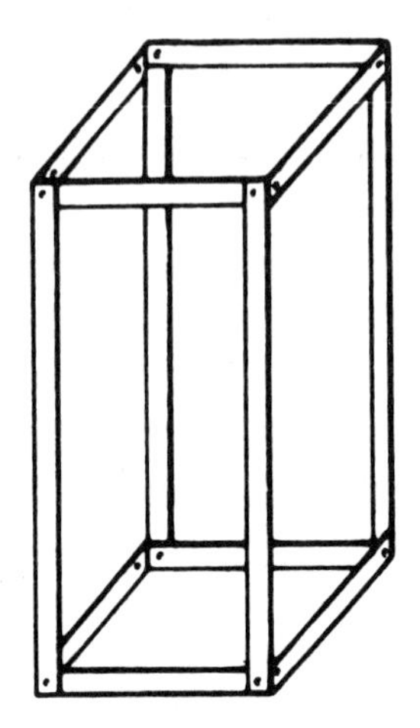

2 Place it in a vertical position as if it were a tower to support a bridge or a crane.

In your notebooks, make a line diagram of the framework you have constructed.

3 Apply a force to the top of your structure by pressing with your hand.

Does the framework support the load applied, without distortion?

If the framework does distort, indicate by means of line diagrams the form of distortion.

4 Analyse these diagrams to see if you can improve the design.

Make sketches of your proposed improvements and write short notes giving reasons for your modifications.

5 Build these improvements into your framework.

6 Re-test the framework to see if it now satisfactorily supports the load.

Record the kind of tests used and the results obtained.

7 ***Are further modifications necessary? If so, make suggestions in your book*** and apply them to your model if you have time.

Structures 3

1 Design and construct a model road bridge using Meccano strips and angle girders of any length. The overall length of the bridge must be at least twice the length of the longest piece of Meccano used. The Meccano members must be joined at their ends. The bridge, when supported at its extremities, must withstand a load, placed anywhere on the roadway, of at least 50 times its own weight. Your teacher will suggest what the minimum length of your model should be, bearing in mind the material available.

(Write the above statement in your notebook as a specification of the project.)

2 ***In your notebooks, make freehand sketches of possible designs and, after studying all your ideas, select the design you think most suitable, giving reason for your choice.***

3 ***Make a larger, more accurate sketch of your final design, and indicate with arrows the position of the load and the position of the supports.*** Analyse the forces acting in each member of the structure, and ***mark with 'C' those members in compression and with 'T' those in tension;*** this will help you make the best choice of material when constructing the bridge.

NOTE: If you are unable to decide if a member is in tension or compression, make no indication on your diagram, but consider it in compression for construction purposes.

4 Construct your bridge. If you find it necessary to modify your original design during construction, ***record these alterations in your notebook, giving reasons.***

5 Place your bridge on the supports provided, and test it by adding, in small increments, a total load at least 50 times greater that the weight of the bridge. Stop loading if distortion takes place, ***noting the maximum load applied.***

6 ***Has your bridge survived the test satisfactorily? If not, attempt to explain the faults, and suggest modifications which should be made.***

1 Design and make a lifting device, powered by an electric motor, suitable for loading the lorry shown with up to 1 kg. The device must be free-standing and be able to lift the required load without toppling over.

(Write the above statement in your notebook as a specification of the project.)

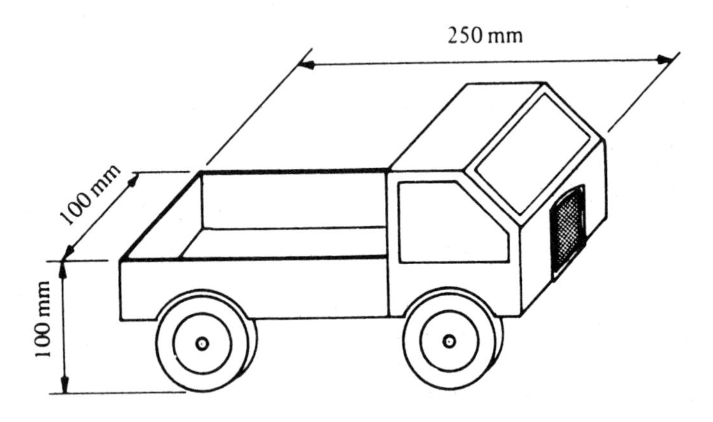

2 There are many ways in which this problem may be solved. You are unlikely to discover the most suitable way immediately. ***Spend several minutes making sketches of various alternative ideas in your notebook.*** (Do not be content with slight modifications to a previous design.)

3 When you have run out of ideas, consider each sketch and decide which provides the most satisfactory solution to the problem. If necessary, produce a new solution by combining ideas from two or more sketches. Aim at using the fewest possible number of members. ***Indicate the design you have selected, and give reasons for your choice.***

4 ***Draw your final design larger and more accurately,*** and analyse the forces acting in the individual members, ***marking with 'C' or 'T' where appropriate.*** This will enable you to make a suitable choice of material during the construction stage.

5 Construct the structural part of your lifting device. If you find it necessary to modify your original design during construction, ***record these alterations in your notebook, giving reasons.***

6 Hang 1 kg from the intended lifting point to establish that:
   i) the structure is sufficiently rigid to support the weight,
   ii) the structure has sufficient stability (i.e. does not fall over).
   Record the result of the test.

7 Fit any pulleys, cord, and hooks which are required, and then bolt on the electric motor — if you are a good designer you will have included a mounting platform in your structure!

8 When complete, test your model thoroughly. Is it entirely satisfactory? Is every member contributing to the strength of the structure? Think about this, and remove any members which may not be required. Is the model still satisfactory?
   Write a short report on how well your model solves the problem, indicating the modifications, if any, that you have made, or those that could be made to improve it still further.

Notes on the electric motor

The motor provided will operate from any d.c. supply within the range 3–12 volts. 12 volts must not be exceeded, or the motor windings will be damaged.

The motor will not operate from an a.c. supply, so ensure that you plug into the correct sockets in the power unit.

A multi-speed gearbox is attached to the motor. You will learn more about this later, but for your lifting device use the 15:1 gear ratio.

You may find it necessary to extend the motor shaft. This may be done by using a coupling and axle — the axle of course must be additionally supported, or the motor shaft will be damaged.

Gears 1

You have been supplied with five Meccano models showing various ways of transmitting rotary motion.

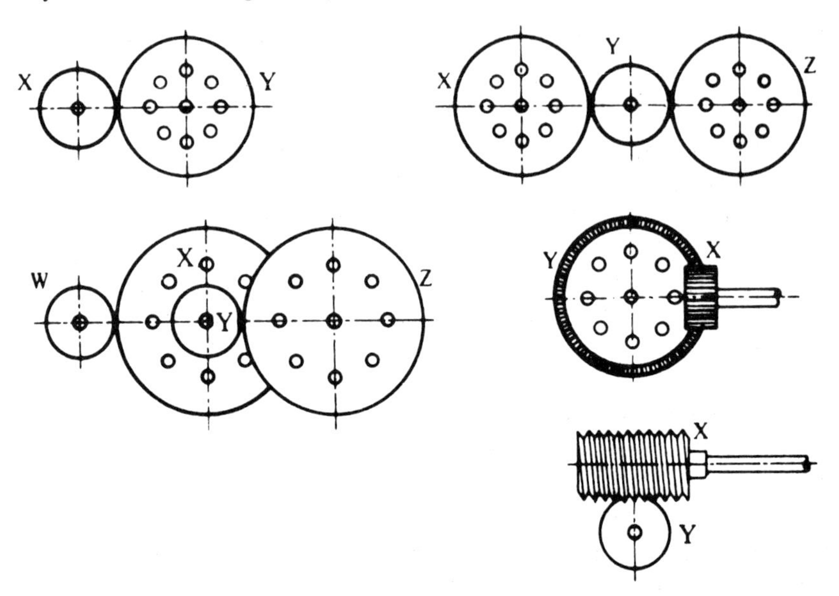

1 ***Make a simple diagram of each gear system in your notebook.*** A gear wheel can be represented by a circle, and gears in mesh can be shown by two circles in contact. Attempt to make the ratio of the diameters of the circles, representing the gears, approximately the same as the ratio of the actual gear diameters.

2 For each of the five gear systems, turn each of the projecting shafts and ***note in your book:***
   i) the number of teeth on each gear;
   ii) the direction in which the gears turn relative to each other — the direction of rotation can be represented thus;
   iii) the number of times the gear shafts turn relative to each other;
   iv) which shaft is easier to turn.

3 When you have investigated each of the five gear systems, study the results carefully and ***make a list in your notebook of similarities and differences between the gear systems.***

Gears 2

These models show other ways of transmitting rotary motion.

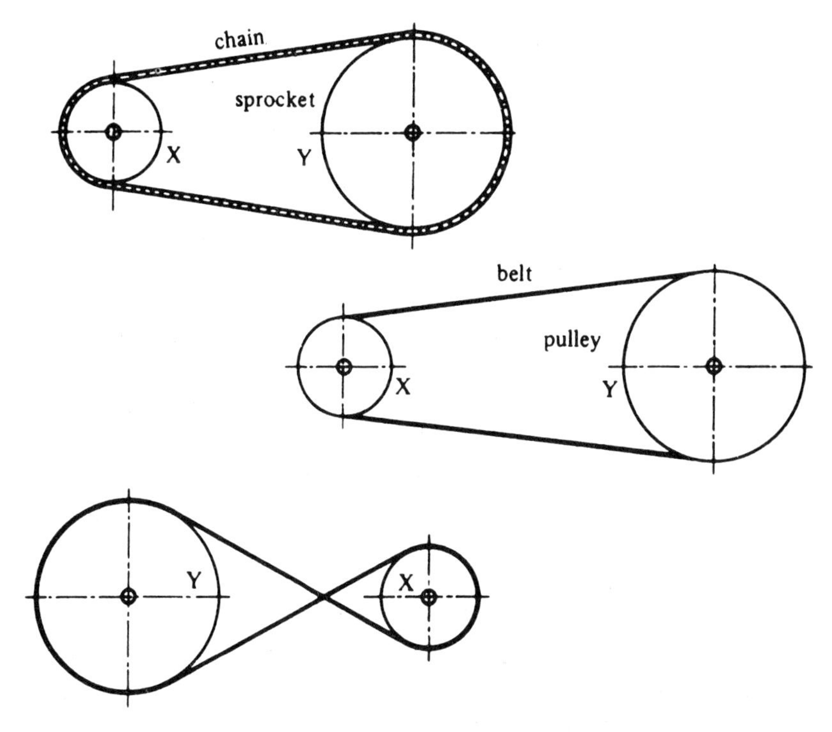

1 *Make simple diagrams of these systems in your notebooks.*

2 For each system, turn each of the projecting shafts and *note in your book:*
   i) the diameters of the pulleys or sprockets;
   ii) the relative direction of rotation;
   iii) the number of times the pulleys or sprockets turn relative to each other;
   iv) which shaft is the easier to turn.

3 *Which system transmits the motion most positively and reliably?*

4 *Summarise your findings in your notebooks.*

5 Compare these systems with those studied in the last lesson. *Note in your books the essential differences.*

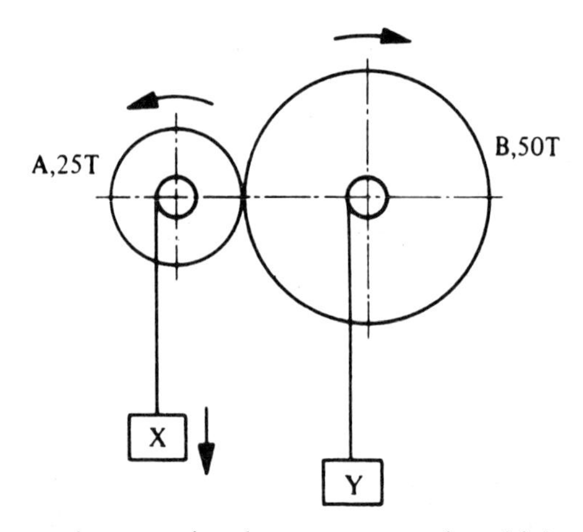

1 The diagram shows a simple gear system in which string is wound around the axle of each gear so that, if one string is pulled down, the other will rise. ***Make a sketch of the system.***

Place a small load on the end of each string, just heavy enough to keep the string taut. ***Note the length of each string from the axle to the load.*** Pull downwards on load X until it has moved a convenient distance. ***Record the final length of each string.***

How far has X been lowered? How far has Y been raised while X has been lowered? Is there a relationship between the ratio of the distances moved and the gear ratio? What is the ratio of the velocity of X to the velocity of Y? (Does X move at the same speed as Y?)

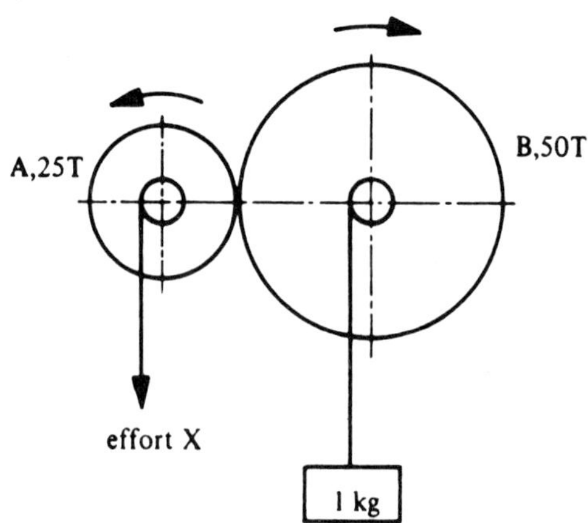

2 i) Using the same system as in 1, make Y 1 kg. What effort (pull force) X do you think will be required to raise the 1 kg? ***Make an estimate, record this in your book, and give your reasons.***

ii) Attach a spring-balance at X and pull on it until the 1 kg is moving with uniform velocity (i.e. the 1 kg moves equal distances in equal times). ***Note the reading on the balance.***

iii) ***Was your estimate recorded in (i) accurate?***

iv) ***What is the ratio of the load to the effort (load/effort)?***

3 i) Repeat the experiment using a similar gear system with nylon-mounted axles. Again attach 1 kg and ***note the effort required to raise the load with uniform velocity.***

ii) ***Was the same effort required in each experiment? If not, suggest possible reasons for this.***

iii) ***What is the ratio of the load to the effort?***

NOTE: The gravitational force acting on 1 kilogram (kg) can be taken as approximately 10 newtons (10 N). The newton is now the preferred unit of force, including load and effort.

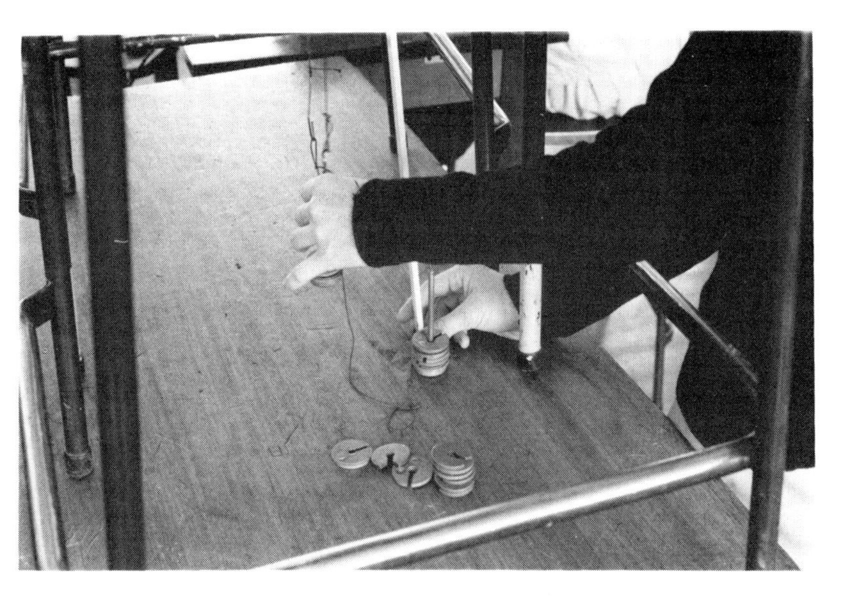

To investigate the effect of the gear ratio on the lifting capability of an electric motor.

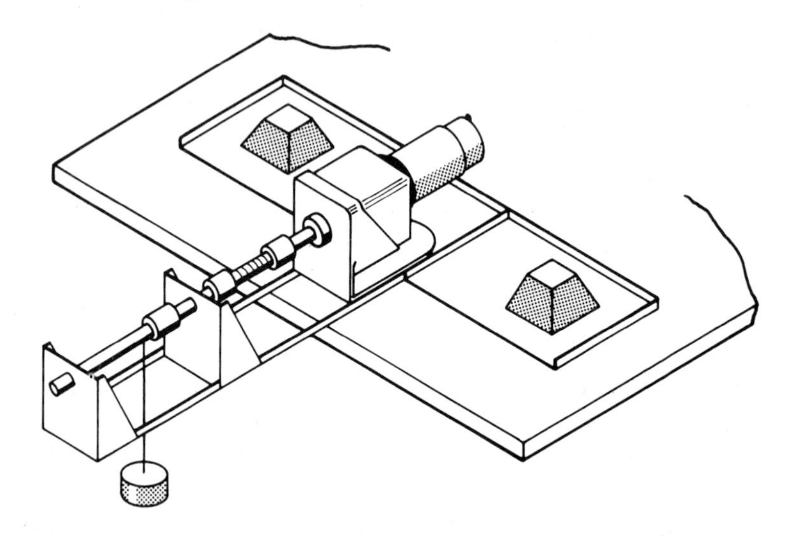

1 Position the lifting device on the edge of a bench so that the string and load hang clear of all obstructions. Place weights on the two rear platforms of the device, to ensure adequate stability during the experiment.

2 Using a 6 V d.c. power supply, connect the circuit shown below.

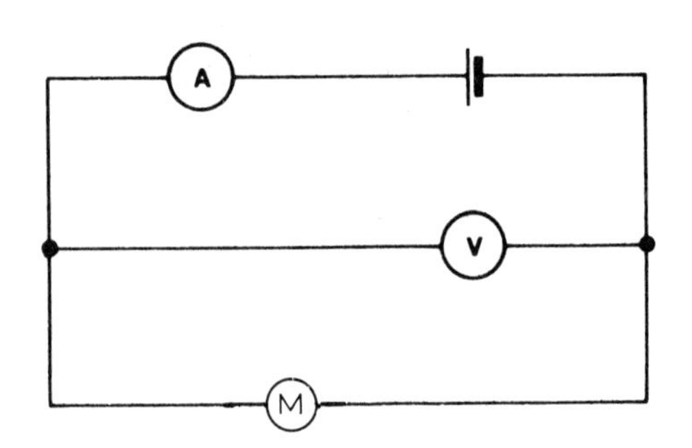

In your notebooks:

i) ***Make a simple line diagram of the apparatus.***

ii) ***Draw the electrical circuit.***

iii) ***Briefly outline the method used.***

iv) ***Draw a graph with the gear ratio on the horizontal axis and the load lifted on the vertical axis.*** (Mark the axis clearly with the appropriate units.)

v) Examine your graph and ***write a full conclusion of your findings from this experiment.***

3 The electric motor you are using is fitted with a four section gearbox. Each section provides a different ratio, 3:1, 4:1, 5:1 or 6:1. They can be assembled using any number, in any order, to give a choice of the following overall gear ratios:
3:1, 4:1, 5:1, 6:1, 12:1, 15:1, 18:1, 20:1, 24:1, 30:1, 60:1, 72:1, 90:1, 120:1, 360:1.

Examples A combination of gearbox sections 5:1 and 6:1 gives an overall gear ratio of $5 \times 6 = 30:1$ In the same way a combination of sections 3:1, 4:1 and 6:1 gives an overall gear ratio of $3 \times 4 \times 6 = 72:1$.

Assemble the units by fitting them end to end. First make sure that the gear units are fitted the correct way round and then fasten all four screws with your fingers. Tighten them *gently* with a small screwdriver.

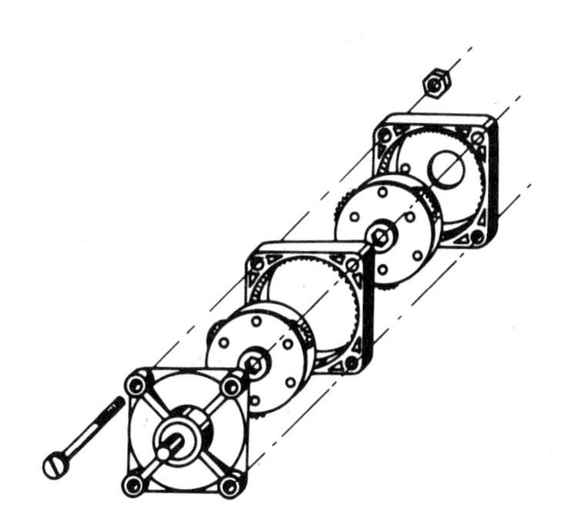

Your teacher may give you a lifting unit with a particular gear ratio provided and marked on the unit. In this case each group in your class will have a different one. You will need to co-operate with other groups so that you use at least five different ratios. Alternatively your teacher may ask you to dismantle and readjust the gearbox in your unit. In this case start with a ratio of 12:1 (a 3:1 section plus a 4:1 section which together give a $3 \times 4 = 12:1$ ratio).

4 Turn on your motor so that the string unwinds and unwind sufficient string to almost reach the floor.

For the first experiment your teacher will recommend a load that is suitable for the gear ratio used. Attach this load to the string. Turn on the motor so that the load is lifted. Adjust your power supply to give an output of approximately 6 volts on the voltmeter and note the ammeter reading whilst lifting this load.

Ensure that the motor is switched off before the load reaches the shaft as the motor will be damaged if the load fouls the shaft.

5 Repeat the 'lift' and note the time taken to lift through 0.5 m.

6 In your note book copy out the following table and record your results for each of the gear ratios.

Voltage (volts)	*Current* (amperes)	*Gear ratio*	*Load lifted*	*Time taken to lift load 0.5 m*
		6:1		
		12:1		
		18:1		
		24:1		
		30:1		

For each of the gear ratios adjust the load on the string until the voltmeter and ammeter readings are the same as they were in your first experiment. (It is important to keep the meter readings the same for all experiments in this assignment.)

NOTE A: Do not use gear ratios below 6:1 because the load will rise so quickly that you will not be able to make adjustments and it is likely that the motor will be damaged.

NOTE B: It is important to make sure that the string only forms a single neat layer on the shaft and does not wind onto previous layers. Why is this important?

Basic Electricity 1

Here are some of the symbols used in electrical circuits:

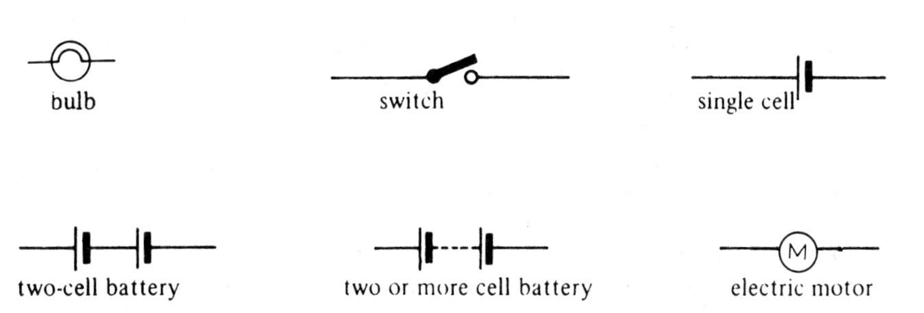

The long line represents the *positive* terminal of the cell and the short line its *negative* terminal.

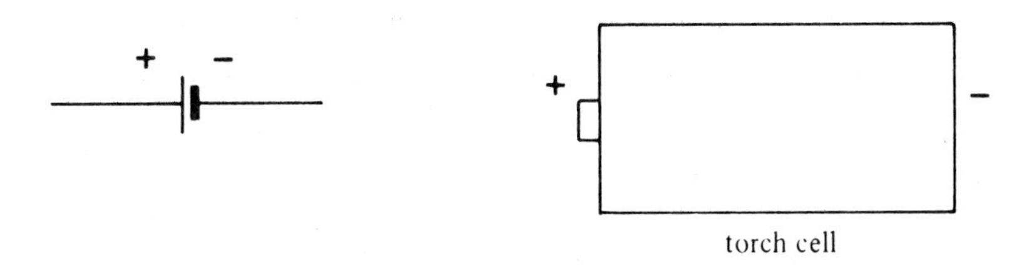

1 Connect a cell to one bulb as shown:

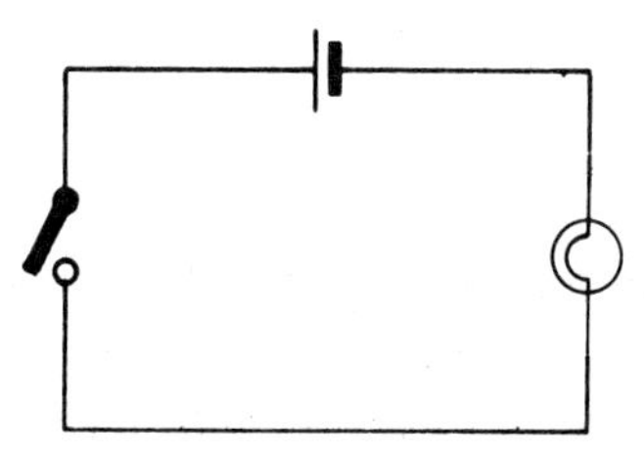

The bulb should light when the switch is closed. Let us call this 'normal' brightness. Reverse the cell and try again. ***What do you conclude?***

2 Connect an ammeter in the circuit. This instrument measures the *rate* at which electricity flows in the wires (conductors). The rate of flow is measured in amperes (symbol A).

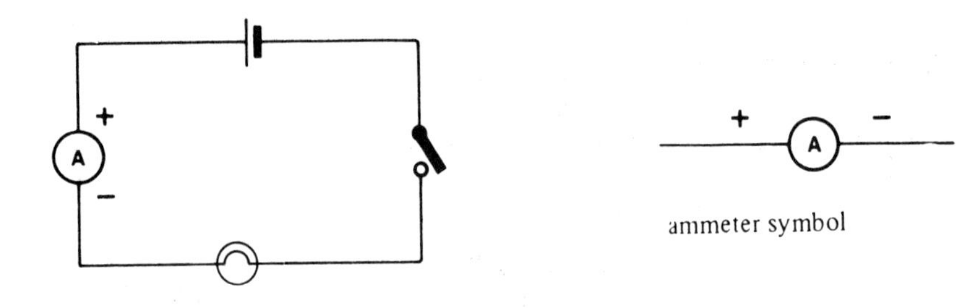

What does the ammeter read?
You have found the current conducted by one bulb when it is used with one cell.

3 Use **two**, then **three** cells, as shown. What does the ammeter read now?

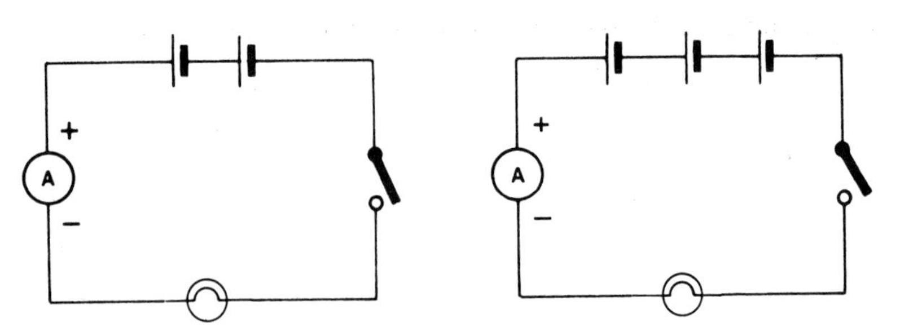

Press the switch for as short a time as possible.

What can you say about the bulb brightness? In each case is it:
a) normal?
b) dim?
c) too bright?
How is the brightness of a bulb related to the current flowing in it?

4 Using one cell, place two bulbs in parallel then two in series.
What can you say about the brightness of the bulb and the current flowing in each case?

How would you measure the current through each bulb in the parallel arrangement?
Try it, and make a note of the currents and the method you used. What can you say about the current through each bulb in the series arrangement?

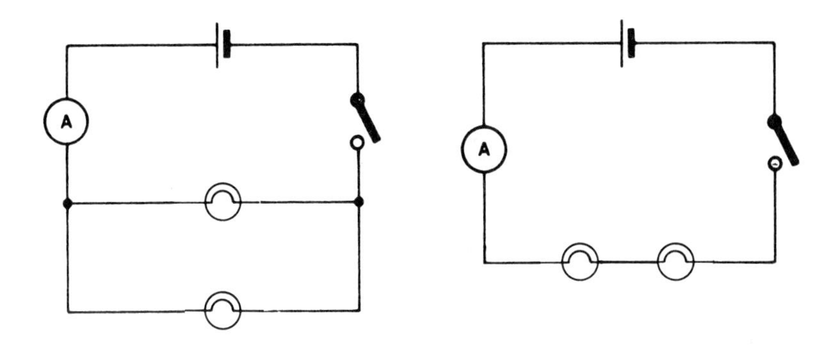

Place the ammeter in different positions in the series circuit and ***make a note of what you find.***

5 Now connect up a circuit with two bulbs in series with two bulbs in parallel. ***Find out all you can about the circuit, and make a record in your notebooks.***

1 (i) Cells provide an electrical force which 'pushes' electrical current round a circuit. If you double the force in a circuit, the current is likely to double. You would probably expect this to be the case in that if you pull on a rope with twice as much force you would expect to be able to lift twice the load.

Connect up a single cell and a bulb as shown, and ***measure the current.***

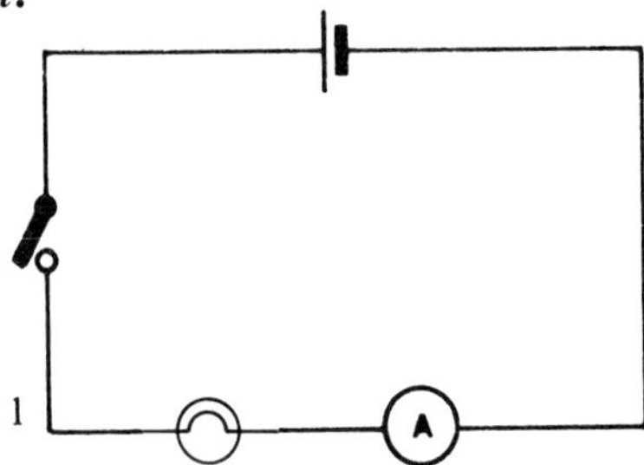

(ii) Place two cells in *series* and ***note the current.***

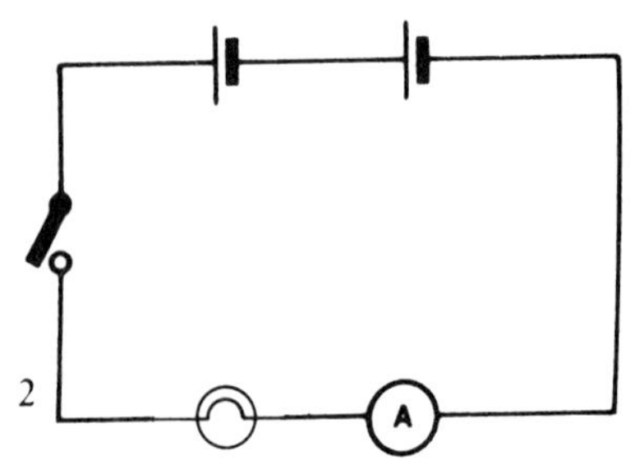

What can you say about the resulting electrical force when two cells are connected in series?

2 A *voltmeter* measures electrical force in *volts* (symbol V) between two terminals of a cell or battery. Use your voltmeter as shown below: place it first across the terminals of one cell, then across the other cell, and finally across both.

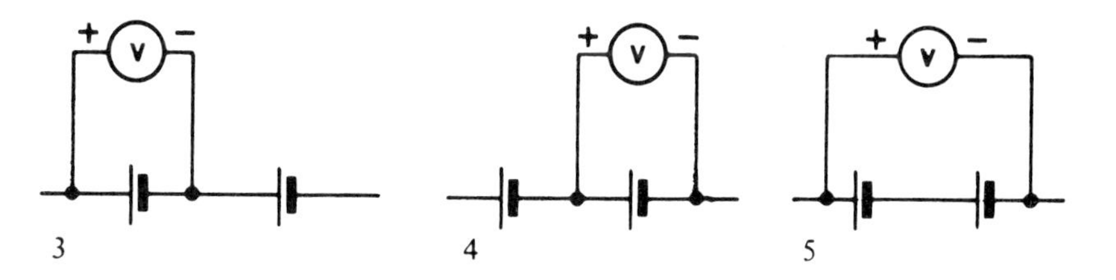

Do your results agree with what was suggested in 1(i)?

3 Set up this circuit, which has two cells in parallel.

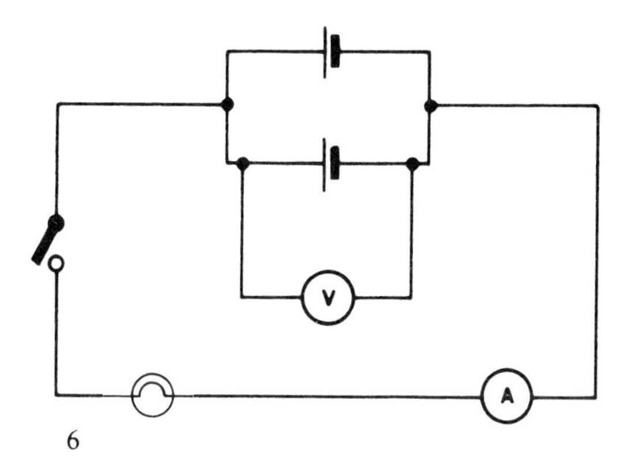

How does your ammeter reading compare with that in circuit 1? Note your voltmeter reading; can you think of a possible reason for connecting cells in parallel?

4 Use your voltmeter to find out whether the cells in a car battery are connected in series or in parallel.

To maintain a current flow in a circuit, an electrical force from a supply such as a cell or battery is needed. Something in the circuit must therefore be resisting current flow. This opposition to current flow is referred to as the *resistance* of the circuit or device within the circuit.

Different materials have different resistances, even if the shape of each piece of material is the same. The best electrical *conductor* among the more common materials is silver; copper is the next best, and then aluminium. Because silver is so expensive, copper is widely used to conduct electricity because of its relatively low resistance compared with other materials.

Current, resistance, and voltage

1 Take a length of resistance wire and apply a sufficient voltage across it to give a current of about 0.2 amperes. ***Measure the current and the voltage.*** (Make sure that the wire does not become hot, just in case this is likely to upset the results at this stage — later on you may wish to make a wire hot on purpose.)

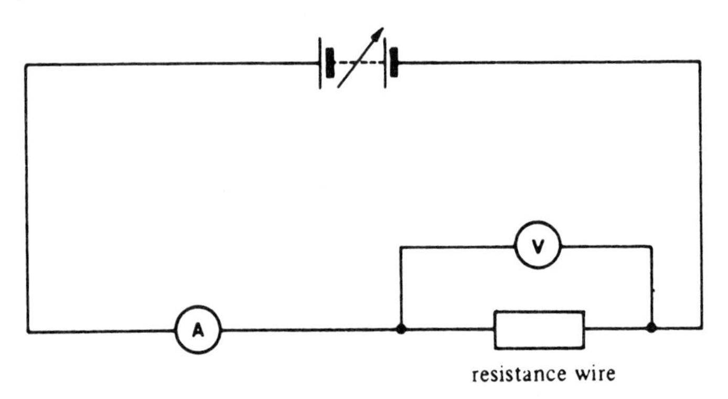

Double the length of resistance wire and ***repeat the experiment.*** By doubling the length of the wire, you have doubled the resistance. Make sure that your supply voltage is set to the same value as before.

What appears to be the relationship between the resistance in a circuit and the current flowing in it, if the electrical force (voltage) remains constant?

You may wish to try different resistance-wire lengths.

2 Take a length of resistance wire, apply a certain voltage across it and ***note the current flowing.*** (Again use a current of about 0.2 A to start with.)

Double the supply voltage. ***What happens to the current?*** Try other voltages but remember that the wire must not become hot.

What appears to be the relationship between the voltage and the current in a circuit if the resistance remains unchanged?

3 ***Try to combine your results from above to make a general rule about the voltage applied to a circuit, the current flowing in it, and the resistance in the circuit.***

NOTE: You may wish to plot a graph of some of your results if you are not entirely satisfied with your conclusions.

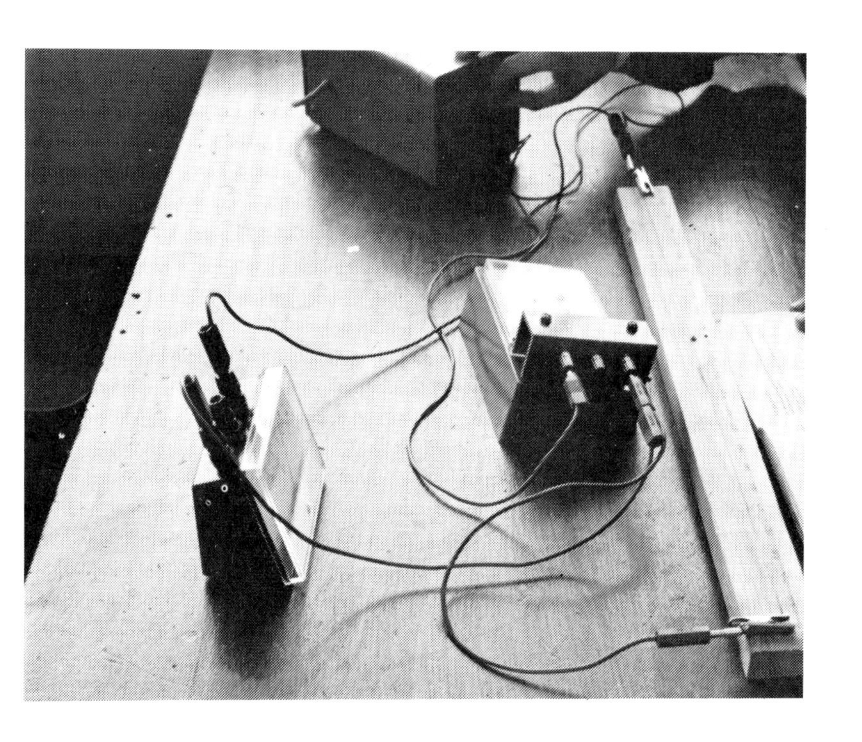

Electrical Switching 1

In your notebook, design a vehicle, propelled by an electric motor, which will run at a speed of about one metre per minute. Do not include a steering mechanism. The diameter of all wheels on the vehicle is 35 mm, and the output shaft of the motor runs at 250 revolutions per minute when operated from a 12 volt supply.

1 In practice it is unlikely that the speed of the output shaft of the motor you have been given will be exactly 250 rev/min. Before you can begin to construct your vehicle, you must know the actual motor speed.

   Devise a method of finding the output shaft speed of your motor, and describe it in your notebook. (Remember that the battery voltage will affect the motor speed.) Now try out your method.

2 It is possible that, owing to a shortage of parts such as wheels and gears, you will have to change your original design.

   Collect together any parts you require.

   What do you predict will be your vehicle speed now that you know the speed of your motor and the exact details of wheel size, gears, etc? (The speed should still be somewhere in the region of 1 m/min.)

3 Construct your vehicle.

4 Try your vehicle, and devise a means of measuring its speed.

   How do your experimental results compare with the answer you have arrived at by calculation in 2 above?

   If they are different, can you suggest a reason?

Electrical Switching 2

1 Connect the circuit as shown and, using your plug-in leads, find out how the switchbox works.

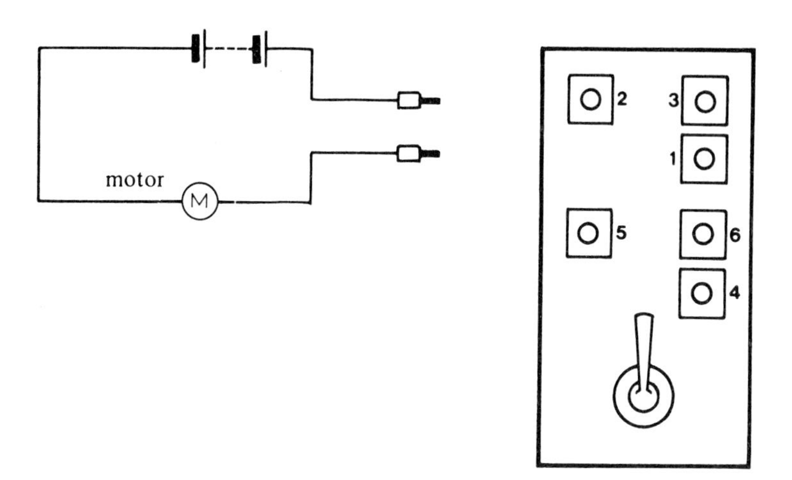

i) Which pairs of sockets switch the motor 'ON' when the switch is 'UP'?

ii) Which pairs of sockets switch the motor 'ON' when the switch is 'DOWN'?

2 Find a way of connecting your switchbox so that both leads to the motor are switched and so that in the 'OFF' position your motor is completely isolated from the power supply. **Care!** It is possible to blow the supply fuse if you make incorrect connections.

Draw your circuit.

What happens if the leads to your supply are interchanged?

Place the motor lead plugs into the yellow sockets. Wire up the remaining four sockets in the box so that with the switch in the 'UP' position the motor runs one way (the vehicle moves forward), and with the switch in the 'DOWN' position it runs the other way (the vehicle moves backwards). ***Draw your circuit.***

Making use of two switchboxes, ***design (on paper) a circuit*** which will start and stop the motor, make it go forward, and make it go in reverse. Try your circuit.

Electrical Switching 3

1. Examine the microswitch which has been provided. Test it to see what kind of switch it is.

   Fit the microswitch so that your vehicle stops if it strikes an obstruction. ***Draw your circuit, using the usual switch symbol for a microswitch. Explain the way your vehicle behaves when using this circuit.***

2. Examine the reed-switch which has been provided. What sort of switch is it?

   Fit the reed-switch and a lamp-indicator unit to your vehicle so that, should the vehicle pass close to a magnet attached to an obstruction, the warning light comes on. ***Draw your circuit*** — the symbol for a normally-open reed-switch is

   Explain how your vehicle behaves when using this circuit.

3. Arrange for your vehicle to *move* with the reed-switch in the motor circuit, then make the vehicle stop when it comes close tc a magnet fitted to an obstruction. The vehicle should restart if the obstruction is removed. ***Draw a diagram of your circuit.***

Electrical Switching 4

Your teacher will show you a number of relays, of various types. It is important that you now understand how a relay works and that you become familiar with the alternative arrangement of contacts.

1 Examine the relay-box provided. It has eight sockets.

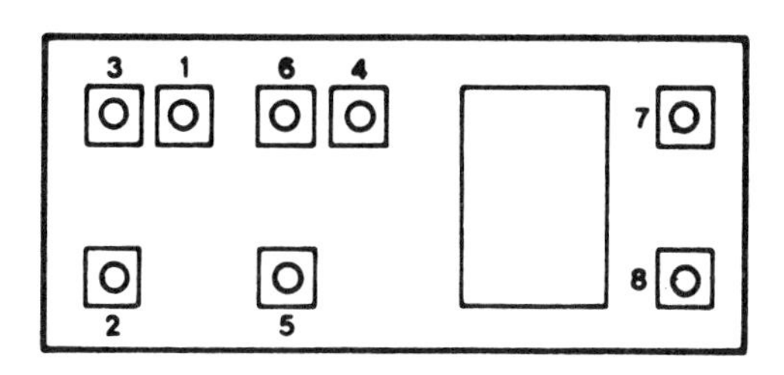

i) The blue and yellow sockets are connected to the relay contacts. ***Find out which pairs of sockets are connected together by the relay contacts when there are no plugs in the green sockets.***

ii) The green sockets are internally connected to the relay coil. Connect these green sockets to a 12 V supply and notice what happens to the relay inside the box when you switch on. ***Now that a current is passing through the coil, find out which pairs of sockets are connected together by the relay contacts.***

iii) ***What kind of switch is being operated by the relay coil?***

2 Fit the relay-box and reed-switch to your vehicle, so that the vehicle stops when it approaches a magnet fitted to an obstruction. *No crashes are allowed.*

Draw your circuit.

3 Now make your vehicle reverse away when its reed-switch approaches the magnet on the obstruction. (Remember your reversing circuits.) ***Any strange effects?***

Observe the operation of the relay as your vehicle moves.

Draw a circuit diagram, if you have succeeded in reversing the vehicle, and write a few notes of explanation.

Electrical Switching 5

In *Electrical switching 4*, you should have fitted a relay-box and reed-switch to your vehicle, and noticed how your vehicle *oscillated* backwards and forwards. If you have not already tried this, do so now.

1 Using the same circuit as the one which produced an oscillating vehicle, replace the reed-switch with a microswitch, and observe the effect when the vehicle strikes an obstruction. Again observe the relay action.

   Draw the circuit in your notebook, and explain what happens.

2 Remove the relay from your vehicle, and disconnect all leads.

   Now connect the relay unit to a 12 volt supply, using the green sockets. Notice what happenes to the relay when you remove one of the plugs which you have placed in the green sockets.

3 Now connect a 5000 μF capacitor between the green sockets. ***Warning!*** The RED plug on the capacitor must be fitted to the green socket connected to the supply POSITIVE. (i.e. observe the polarity of the capacitor.)

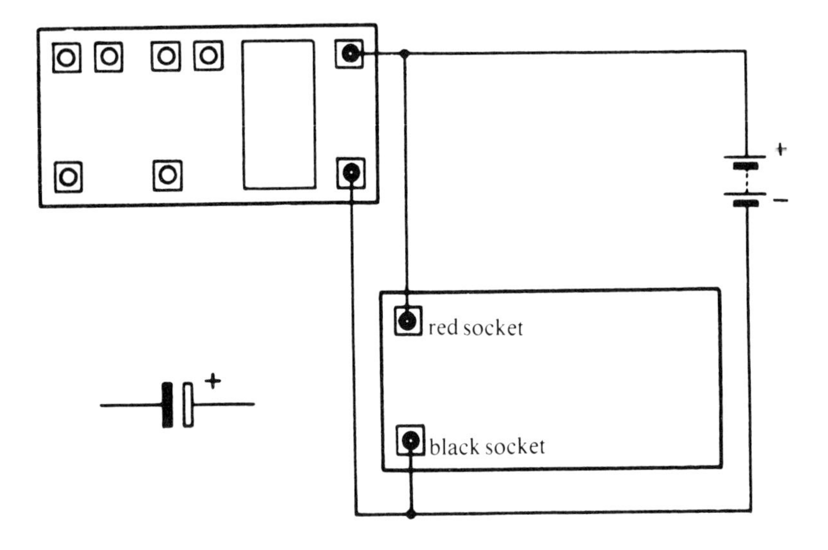

Note what happens to the relay when you remove one of the leads to the supply. Try to think of an explanation of what you see happening.

Draw the layout and circuit diagram in your notebook. Explain what happens.

4 Now refit the relay and microswitch to your vehicle. Connect up the reversing circuit which you used in 2 above, but this time fit a capacitor (approximately 2000 μF) across the relay coil. (Again observe the polarity of the capacitor.)

Make a note of what happens when the vehicle strikes an obstruction.

5 ***Measure how far the vehicle reverses from the obstruction.***

Fit a second capacitor (again approximately 2000 μF) in parallel with the first. Now replace the parallel pair of capacitors with a single capacitor of value approximately twice the value of the one you first used. ***In each case, measure how far the vehicle reverses from the obstruction.***

Now try fitting two of these larger capacitors in series across the relay coil. (Connect the positive end of one capacitor to the negative end of the other; the free ends should be connected across the relay coil.)

How far has the vehicle reversed this time?

6 ***What conclusions can you draw from your investigations in 5?***

Electrical Switching 6

1 Examine the double-relay unit you have been given. The yellow and blue sockets are the relay contacts which you can use to operate other equipment, such as a motor. Notice that two blue sockets and one yellow are positioned next to one relay, and a similar set of three sockets is positioned next to the other relay. ***Observe what happens in the following cases, and note this down.***

2 Connect a 12 V d.c. supply to the pair of red sockets. Leave the supply connected for the following investigations.

3 Short-circuit one pair of green sockets, then remove the short. Again short-circuit the *same* pair of green sockets. Remove the short-circuit.

4 Short-circuit the other pair of green sockets. Remove the short. Again apply a short-circuit to the same pair.

5 Find out which of the yellow and blue sockets become joined by relay contacts as each relay energies and de-energises. ***Copy out and complete the table below, making use of the diagram.***

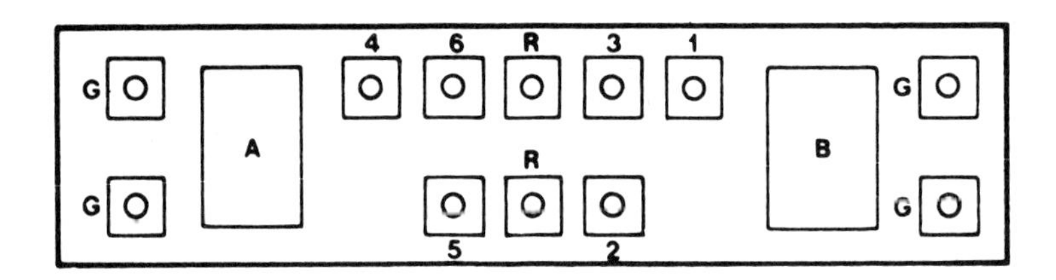

	sockets joined together
Relay A energised Relay B de-energised	
Relay A de-energised Relay B energised	

6 Use the unit to make your vehicle reverse when it strikes an obstruction. On striking a second obstruction, make it again reverse towards the first obstruction. The relay unit should be wired up on the vehicle itself so that one pair of supply leads leaves the vehicle.

The circuit of the double-relay unit is:

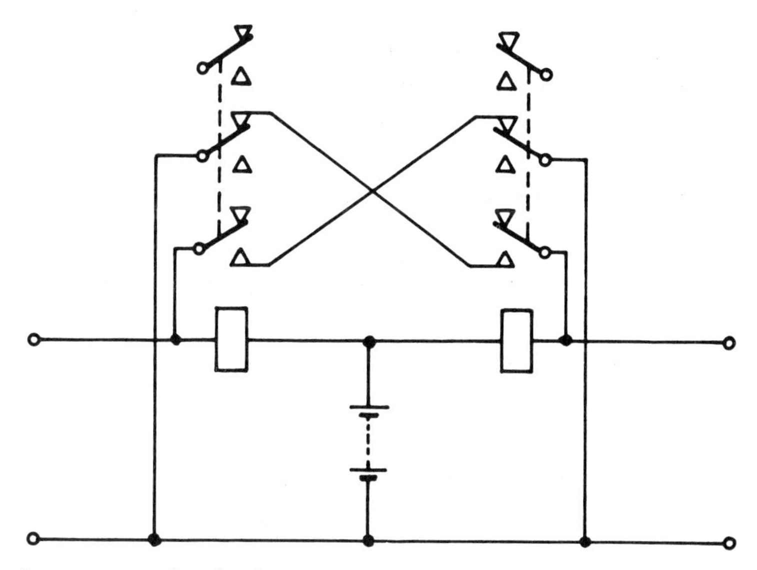

In your notebook, draw:

i) ***a wiring layout of the circuit you have used on your vehicle,***

ii) ***a circuit diagram which includes the above circuit of the bistable relay unit.***

7 From the circuit in 6, try to work out how the relay unit works.

1 You will have been given a photocell, type ORP 12, mounted within a tube, a 12 V light source, complete with tube and lens, and also a lamp indicator unit. A sensitive indicator lamp is needed for this assignment. Your teacher will give you a suitable bulb.

Connect up the following circuit.

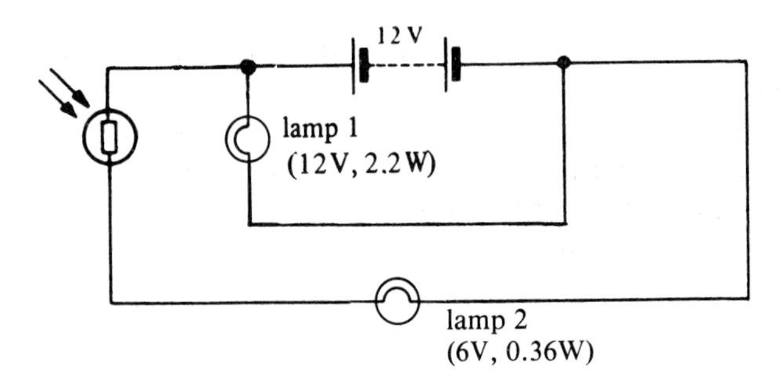

i) Approximately align the light source and the photocell so that there is a space of about 150 mm between the ends of the tubes.

Now make alignment adjustments to both the light source and the photocell holder until the maximum illumination of the photocell is obtained.

Warning: If too high a current passes in the photocell, there is a danger that it will 'burn out'. This is possible under certain lighting conditions where the photocell is fully illuminated. For example, if there is only a bulb and a photocell in series across a 12 V supply, the cell will 'burn out' if it is placed less than about 150 mm from the light source *for a long period of time.* At other distances, long periods of use should be avoided.

There will be *no* danger of the cell 'burning out' if the resistance of a device in series with it is greater than about 200 ohms. For example, the relay coil in the relay unit has a resistance of 185 ohms, and would be quite safe. On the other hand, a 6 V, 0.36 watt bulb, used in the indicator unit, has a resistance of only about 10 ohms when cold and about 100 ohms when hot. Therefore care should be taken not to leave the apparatus switched on for long periods of time.

What happens to lamp 2?

ii) Move the light source further away from the photocell, again checking the alignment. ***What happens to lamp 2?***

iii) Place your hand in the light beam. ***What happens to lamp 2? Give, as clearly as you can, an explanation for your results.***

2 Replace lamp 2 with a relay-box.

i) ***Is the relay energised or de-energised when light is hitting the cell?***

ii) Place your hand in the light beam. ***What happens to the relay?***

iii) Increase the distance between the photocell and the light source, and ***note the greatest spacing possible to ensure reliable operation of the relay.***

iv) Connect your vehicle motor to the output sockets of the relay. Use this arrangement to stop your vehicle when it breaks a beam of light.

Draw your circuit.

Electrical Switching 8

1 In the previous assignment you designed a circuit using a photocell, a light source, and a relay unit to make your vehicle stop when the light beam to the photocell was interrupted.

Modify this circuit to make your vehicle oscillate rapidly in and out of the light beam, making use of a reversing circuit.
Draw the circuit in your notebook.

2 i) Using the same circuit, add a capacitor of approximately 2000 μF to 'delay' the operation of the relay.
Direct your vehicle towards the light beam, and ***record what happens.***

ii) Increase the delay capacitor to a value approximately double that used in (i), and repeat the experiment, ***recording the behaviour of the vehicle.***

iii) Increase the capacity further, to a total of say 10 000 μF, and repeat the experiment. ***Note the effect on the vehicle.***

3 ***Did the vehicle behave as you had expected?***
Explain, as fully as you can, the operation of this circuit.

Electrical Switching 9

Now that you understand the operation of the circuit used in the previous assignments, ***design in your book a circuit*** using a light source, a photocell, 2 relay units (do you remember the purpose of the second relay?), and a capacitor, to make your vehicle reverse for a short distance when it interrupts the beam of light shining into the photocell.

Electrical Switching 10

1 Do you remember how to operate the bistable relay unit? A short-circuit across one pair of green sockets causes the adjacent relay to energise. Change-over is achieved by *removing* this short and applying a short across the other pair of green sockets.

i) What will happen if illuminated photocells are connected across each of the pairs of green sockets? ***Try it, and record what happens.***

Blank off each photocell in turn, and then both together. ***Does the bistable work normally?***

ii) Blank off both photocells. Now allow one photocell to be illuminated. Blank this off again, and illuminate the other photocell. Blank off both, then illuminate the first cell again.

Record what happens in your notebook.

Does the bistable work normally?

2 By introducing two additional single-relay units, one for each photocell, modify the previous circuit to give complete control of the bistable when both photocells are normally illuminated.

Did you succeed? If so, ***draw the circuit in your notebook.***

3 Now ***design in your notebook*** a circuit which will make your vehicle move backwards and forwards between two illuminated photocells. Connect up the circuit and try it.

Linear Motion 1

1 So far in the course, most of the devices which you have constructed have been made to operate by using an electric motor. The output shaft of an electric motor rotates, i.e. it has rotary motion. You will have seen machines or devices where certain parts move linearly (in a straight line). How many different ways can you think of which would enable rotary motion to be converted into linear motion?

Draw diagrams of a number of possible methods in your notebook, giving each a suitable title. Try to make the methods as different as possible.

2 Choose one of your designs and construct it from Meccano. Make provision for fixing an electric motor but, before fitting one, turn the device by hand to produce the rotary motion. Estimate how much force can be exerted by the part of the device which moves linearly.

What factors determine the force that your mechanisms can exert?

3 Fit the electric motor.

i) ***How much linear movement is provided by your device, and how can this distance be varied?***

ii) ***Will your device push and pull equally well?***

4 Place 1 kg on the bench. Arrange for your device to produce slow linear motion. Will it push the 1 kg along the bench against the force of friction between the 1 kg and the bench?

5 ***Make notes in your notebook, explaining where your design has succeeded and where it has failed.***

Linear Motion 2

1 Electrically operated solenoids and air-operated pneumatic cylinders are convenient devices for obtaining linear motion directly, rather than by converting rotary motion to linear motion. You will have been given some examples of solenoids and pneumatic cylinders; examine each to familiarise yourself with its construction and appearance.

2 Connect an on-off valve and air supply to:
   i) a double-acting pneumatic cylinder, and
   ii) a single-acting pneumatic cylinder.

   Make notes on the actions performed.

3 ***Make a list of the factors which you think will affect the force provided by the piston of a pneumatic cylinder.***

4 Connect a solenoid to an on-off switch and a 12 volt d.c. supply.

   Note what happens to the moving part of the solenoid (the armature). Note any effects which you think are particularly important.

5 ***Make a list of the factors which you think will affect the force provided by a solenoid.***

6 The next set of assignments suggests some investigations which you can make to discover some important properties of pneumatic cylinders and solenoids.

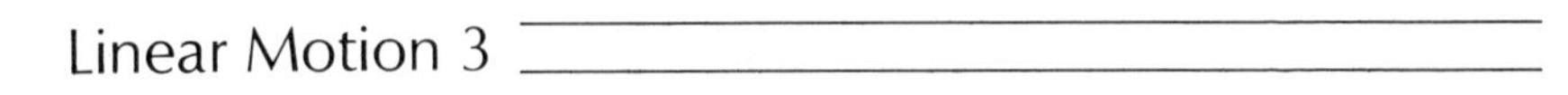

Preliminary notes

1 You will already have learned that certain factors may affect the force produced by an electrically operated solenoid or by the piston of a pneumatic cylinder.

These factors include:

for the solenoid

i) the length of armature inside the solenoid,
ii) the current flowing in the solenoid coil,
iii) the number of turns in the solenoid coil,

for the pneumatic cylinder

i) the position of the piston in the cylinder,
ii) the pressure of the air supply,
iii) the area of cross-section of the piston,
iv) if the cylinder is double-acting, the direction in which the piston is moving, i.e. whether it is pushing or pulling.

Before attempting ANY experiment, all possible variable quantities or values must be kept constant, with the exception of two. Of these two, we vary one and observe what happens to the other, remembering that THROUGHOUT the experiment all other conditions must be kept as constant as possible.

For example, if you wish to investigate how the *pressure of the air supply* affects the *force* produced by a pneumatic cylinder, you must:

i) keep the position of the piston constant;
ii) use the same cylinder throughout (i.e. keep the area of cross-section constant);
iii) feed air into the same end of the cylinder throughout the experiment (i.e. the cylinder must push (or pull) throughout the experiment).

In experiments with pneumatic cylinders or solenoids, however, you will find that keeping armature and piston positions constant could slow down your investigations. You would be wise, therefore, to first investigate how the position of an armature (or piston) affects the force produced. Then, if you find that position does not affect the force, there is no need to keep the position fixed.

2 Work through the following investigations, some of which may be demonstrated by your teacher. You may attempt the assignments in any order, but remember the hint given in the last paragraph. After each investigation, record your results neatly in your book and give *as many conclusions as possible*. (After all, you are making investigations to arrive at conclusions.)

3 Refer to the notes on *The use of graphs*. Read through these to find out what information on graphs is available to you. (After you have plotted graphs during investigations, you will find the notes more useful and meaningful than at present.)

4 Before writing up your experiments, refer to the notes on *The approach to investigational experiments*.

3.1 Solenoids

To investigate the relationship between the position of a solenoid armature and the force produced

1 Set up the apparatus as shown in the diagram using the solenoid connections labelled 'start' and 1000 turns'.

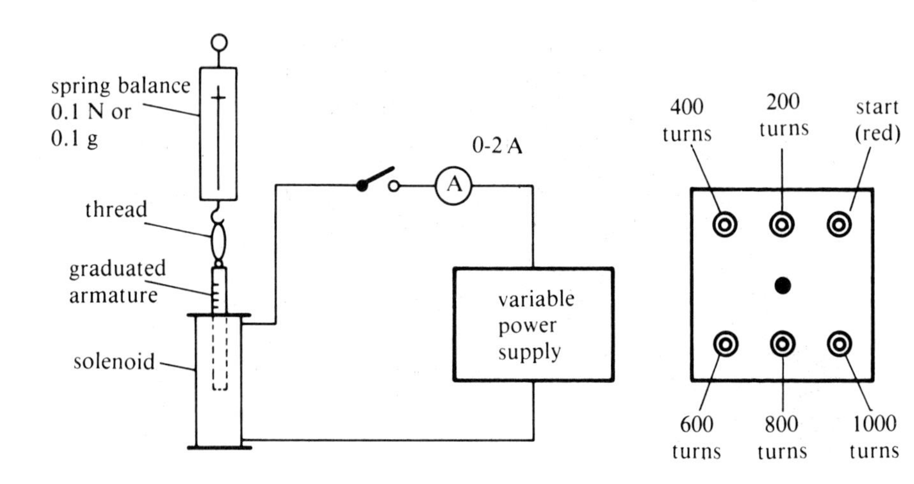

Support the spring-balance at S, using one clamp of a retort stand, and secure the coil rigidly using a second clamp on thc same retort stand.

2 ***Take a spring-balance reading***, with no current flowing in the coil.

3 Switch on the variable-voltage power supply and adjust it to provide a current one ampere.

4 Adjust the spring-balance support (S) until the graduated armature is just inside the coil. ***Note the spring-balance reading and armature position.*** Switch off; do not leave the current flowing too long. ***What will happen if you do? There are a number of possibilities.***

5 ***Make a table of your results*** — remember that the only force you are concerned with is that produced between the armature and the solenoid when the latter is passing a current.

6 ***Repeat 4 for different lengths of armature inside the solenoid.***

7 ***Draw a graph from your results.*** Refer to *The use of graphs*, which will suggest what you should plot on the x and y axes.

8 ***What conclusions do you draw from the graph?***

NOTE: At the end of your written report, make notes on anything which you consider to be important and which you have not mentioned elsewhere in the report.

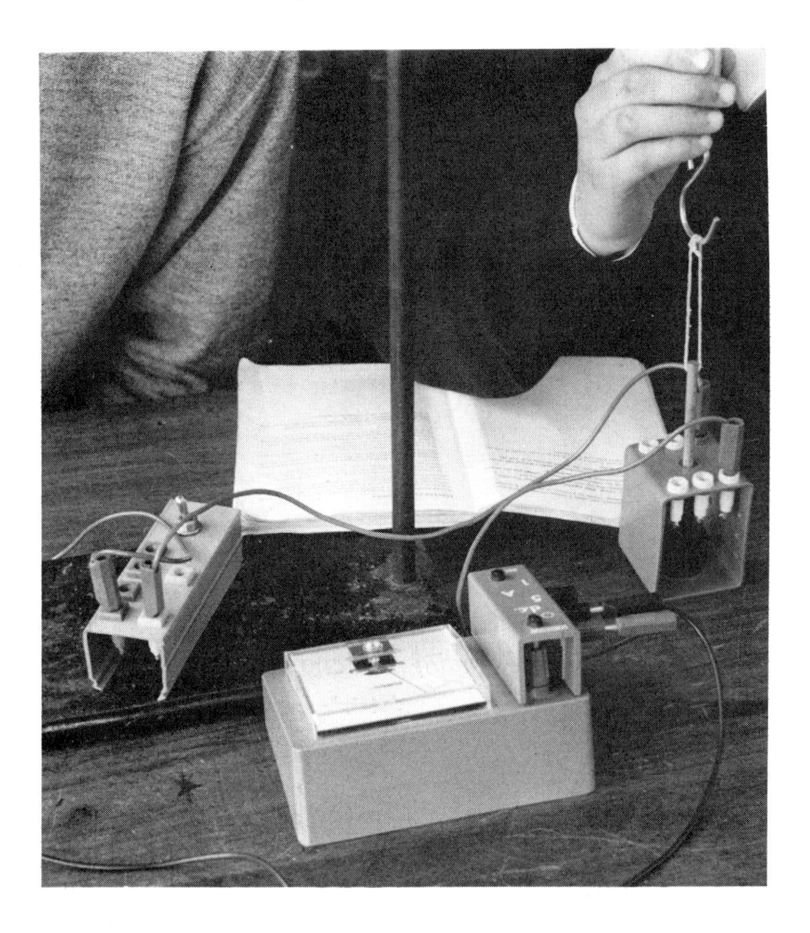

3.2 Solenoids

To investigate the relationship between the current flowing in a solenoid and the force produced

1 Set up the apparatus as shown in the diagram, using the solenoid connections labelled 'start' and '1000 turns'.

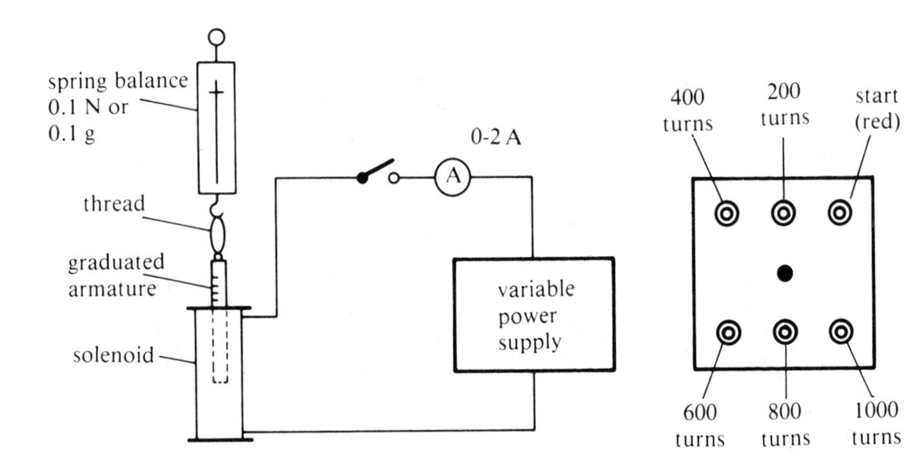

Support the spring-balance at S, using one clamp of a retort stand, and secure the coil rigidly using a second clamp on the same retort stand.

2 ***Take a spring-balance reading with no current flowing in the coil.***

3 Switch on the supply and find the *maximum* current you can allow to flow in the coil without it getting very hot. Do not exceed 80°C (which is about the temperature of a hot radiator).

4 ***Make a table of your results*** — remember that the only force you are concerned with is that produced between the armature and the coil when the latter is passing a current.

5 *Take readings of the spring-balance for different currents flowing in the coil, up to the maximum you have found in 3.* Remember that, for each value of current, **the length of armature inside the coil may need to be kept constant**, unless you found this precaution was unnecessary in assignment 3.1.

6 ***If necessary, draw a graph from your results.*** Refer to *The use of graphs*, which will help you to draw and interpret your graph.

7 ***What conclusions do you draw from the experiment?***

NOTE: At the end of your written report, make notes on anything which you consider is important and which you have not mentioned elsewhere in the report.

3.3 Solenoids

To investigate the relationship between the number of turns on a solenoid and the forces produced

1 Set up the apparatus as shown in the diagram, starting with a coil of about 1000 turns.

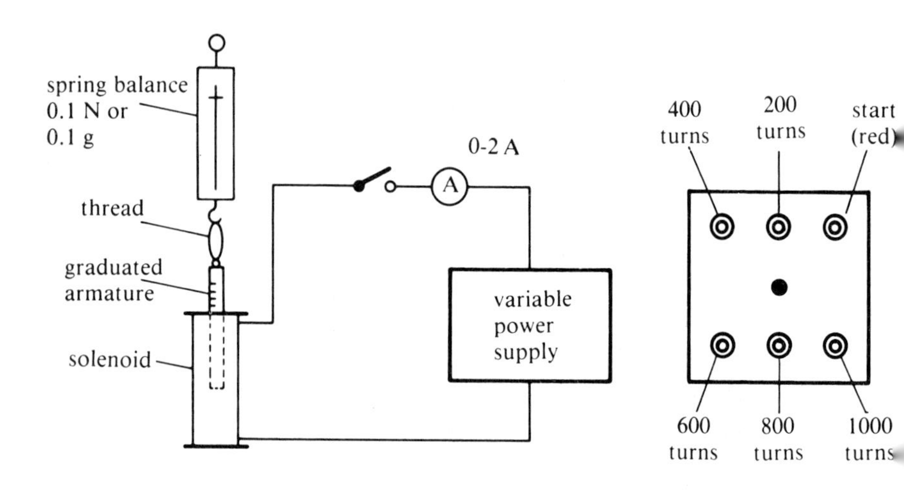

Support the spring balance at S, using one clamp of a retort stand, and secure the coils rigidly using a second clamp on the same retort stand.

2 ***Take a spring-balance reading with no current flowing in the coil.***

3 Switch on the supply and adjust the current to a value of about 1 ampere.

4 Adjust the position of the armature until a suitable high reading on the spring-balance is obtained. (The length of armature inside the coil should not exceed about 25 mm). ***Note the spring-balance reading, the length of armature inside the solenoid, and the current.***

5 *Make a table for your results* — remember that the only force you are concerned with is that produced between the armature and the coil when the latter is passing a current.

6 Repeat 4 using coils with various numbers of turns. For each, check that the current and armature position are constant, unless you have already performed experiments which show that these precautions are unnecessary. ***Note the number of turns and the spring-balance reading each time.***

7 ***If necessary, draw a graph from your results.*** Refer to *The use of graphs*, which will help you to draw and interpret your graph.

8 ***What conclusions do you draw from the experiment?***

NOTE: At the end of your written report, make notes on anything you consider is important and which you have not mentioned elsewhere in the report.

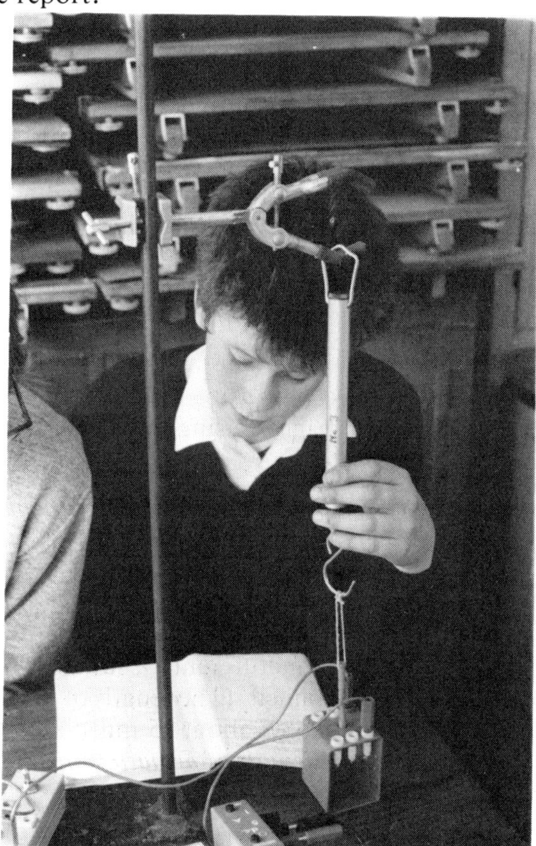

Examine the following components which are required for the next investigations.

Pressure reducing valve with gauge (Regulator)

Symbol

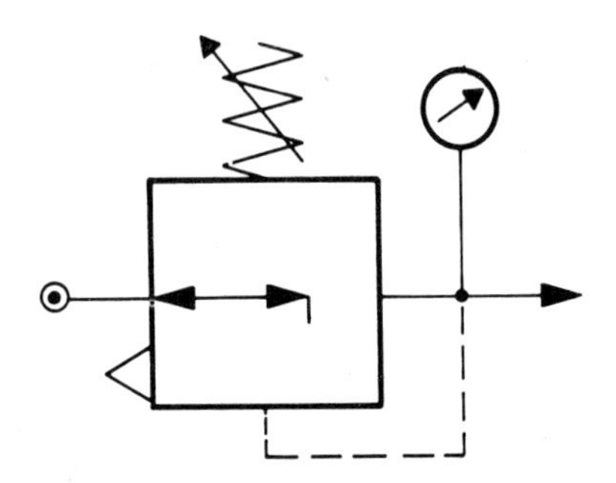

Your teacher will explain and demonstrate the functions of the pressure reducing valve. It reduces the air supply from the compressor/reservoir to a pressure which is suitable for working through your assignments.

It is important that you observe the safety precautions associated with pneumatic equipment, explained and demonstrated by your teacher.

The pressure reducing valve is the first component in the air-supply from the compressor/reservoir.

Push button three port valve

The standard symbol for this two-state ('on/off') type of valve is shown below. The two squares which help to make up the complete symbol are used to represent these two possible conditions.

In the top diagram left half of the symbol represents the valve in the 'on', or 'operated' state. (This will apply irrespective of the method of operation — push-button, roller trip, etc.) The other square represents the valve in the 'off', or 'un-operated', state, as seen in the diagram at the bottom.

It should be noted that in all circuit diagrams the complete symbol is drawn but in most cases the pipe line connections will be drawn to positions on the outside of the lower square — seen here — the 'un-operated' condition of the valve.

Operated

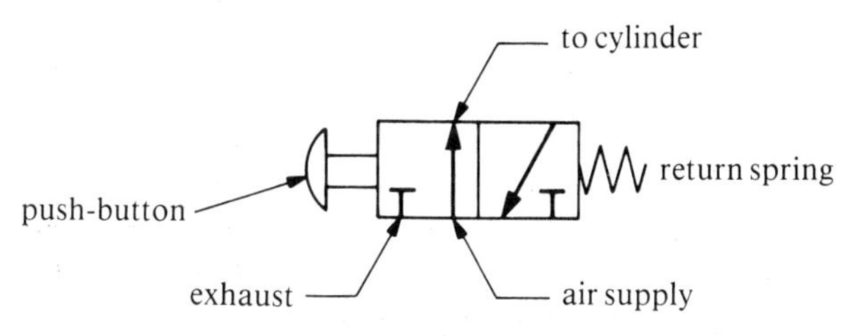

Unoperated

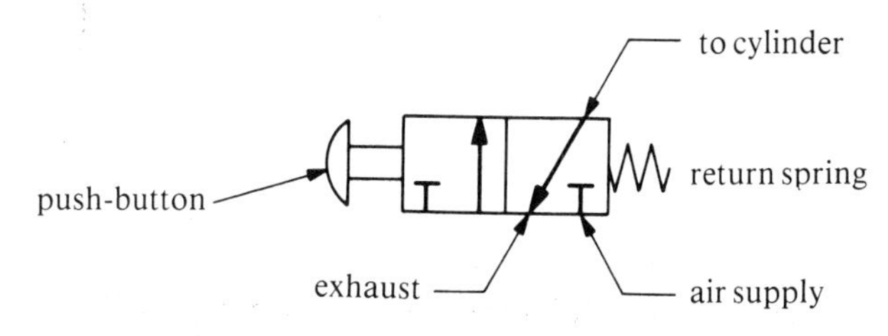

In this type of valve, the air supply must always be connected to the inlet port. (If incorrectly connected, the valve will not function properly because the ball, which is used to provide an air-right seal, will be forced away from its seating and leakage will occur.)

Sectional view of three port 'Poppet' valve.

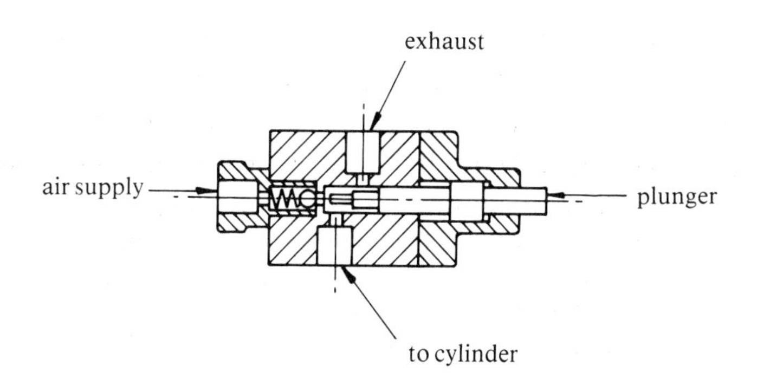

The single-acting (S/A) cylinder

This has one pipe connection and one exhaust aperture. A sectional view is shown below together with the symbol.

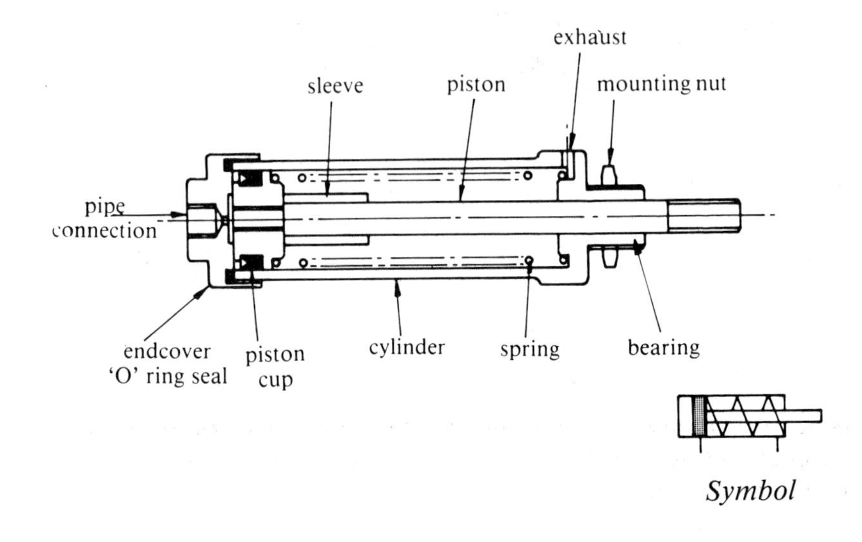

Symbol

Properties

This type of cylinder produces large push forces only. The retraction of the piston rod is by an internally mounted spring and therefore this movement is not usable as a power stroke.

The double-acting (D/A) cylinder

This type of cylinder has two connections, one at each end, and the piston rod can be moved quite easily when neither of these is connected to an air supply.

A sectional view is shown on page 45 together with the symbol.

Properties

This type of cylinder produces both powerful push and pull forces.

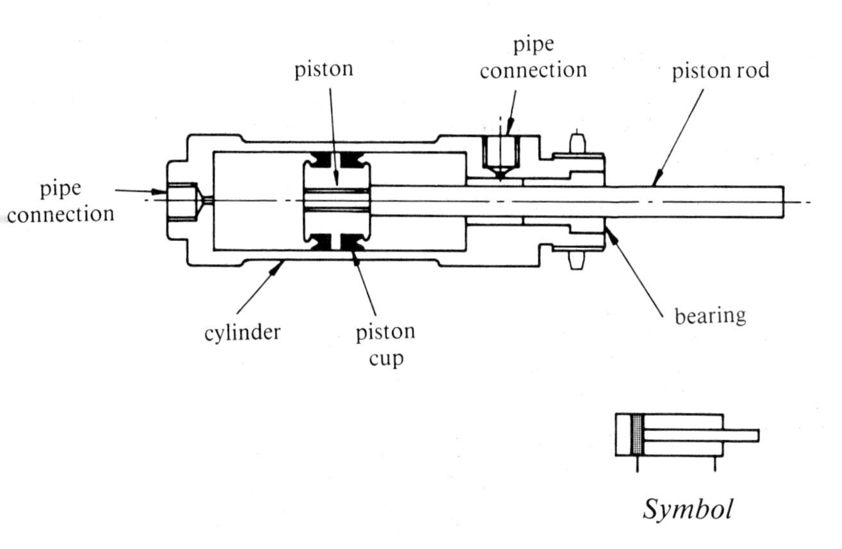

Symbol

Sketch the standard symbol for the three port valve and for both the single acting and the double acting cylinder in your notebook for future references.

3.4 Pneumatic Cylinders

To investigate the relationship between the position of the piston and the force produced

1 Set up the frame and cylinder as shown.

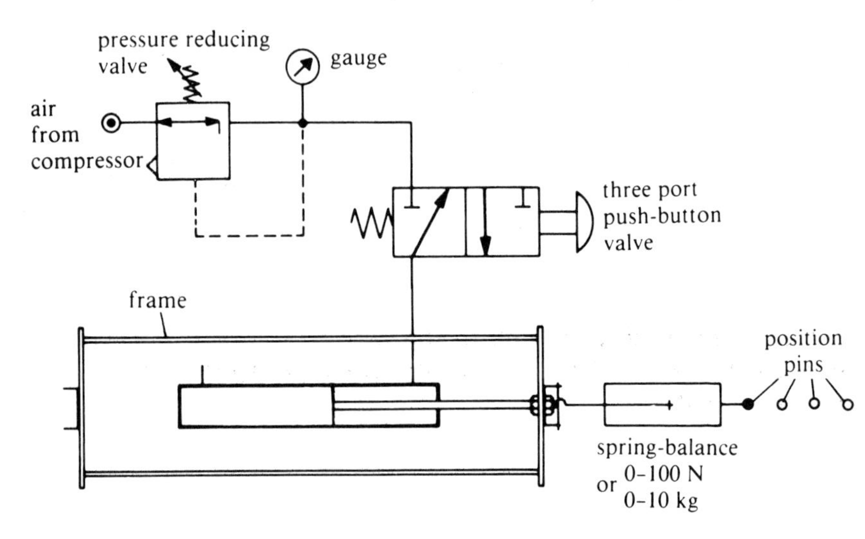

2 Adjust the position of the spring-balance so that it reads zero with the piston fully out.

3 Connect the air supply to inlet Y, and include an on-off valve and a pressure-reducing valve in the line.

4 Operate the on-off valve, and adjust the pressure on the pressure-reducing valve to some definite value to give a reading on the spring-balance between half and full-scale value.

5 Note the spring-balance reading and the length of the piston roc projecting from the cylinder.

6 Repeat 5 for different positions of the piston rod by adjusting the position of the spring-balance. ***Note the spring-balance reading and the length of piston rod projecting from the cylinder each time.***

Make a table from your results. Include only the variable quantities in the table headings.

7 ***If necessary, draw a graph from your results.*** Refer to *The use of graphs*, which will help you to draw and interpret your graph.

8 ***What conclusions do you draw from your experiment?***

NOTE: At the end of your written report, make notes on anything which you consider is important and which you have not mentioned elsewhere in the report.

3.5 Pneumatic Cylinders

To investigate the relationship between the pressure of the air supply and the force produced

1 Set up the apparatus as shown.

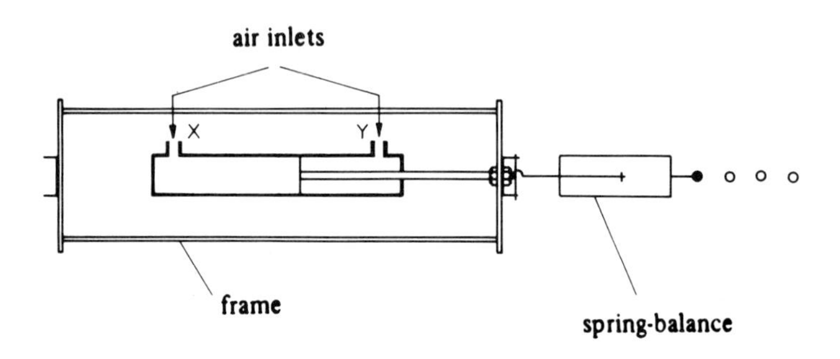

2 Adjust the position of the spring-balance so that it reads zero with the piston fully out.

3 Connect the air supply to inlet Y, including an on-off valve and a pressure-reducing valve in the line. Ensure that the air supply is shut off at both the on-off and the pressure-regulating valves.

4 Switch on the air supply, and adjust the pressure-reducing valve until a noticeable reading is obtained on the spring-balance. ***Make a table for your results, and note the spring-balance and pressure readings.***

5 ***Repeat 4, noting the force produced at different air pressures.*** (HINT: For reliable results, do not at any time *reduce* the air pressure by small amounts before taking a reading. If you increase the pressure more than you intended, turn down the pressure to zero and begin to increase the pressure again.)

6 ***If necessary, draw a graph from your results.*** Refer to *The use of graphs*, which will help you to draw and interpret your graph.

7 Repeat the above experiment, but fit the spring-balance to the opposite end of the frame, and supply air to inlet X instead of

inlet Y. This time you are investigating the piston rod 'push forces', rather than 'pull forces'.

8 ***Draw a second graph superimposed upon the previous one.***

9 Compare the two graphs.

10 ***What conclusions do you draw from your experiments?***

NOTE: At the end of your written report, make notes on anything which you consider is important and which you have not mentioned elsewhere in your report.

When you have performed this assignment, attempt *assignment 3.6.*

3.6 Pneumatic Cylinders

1 Set up the apparatus as shown.

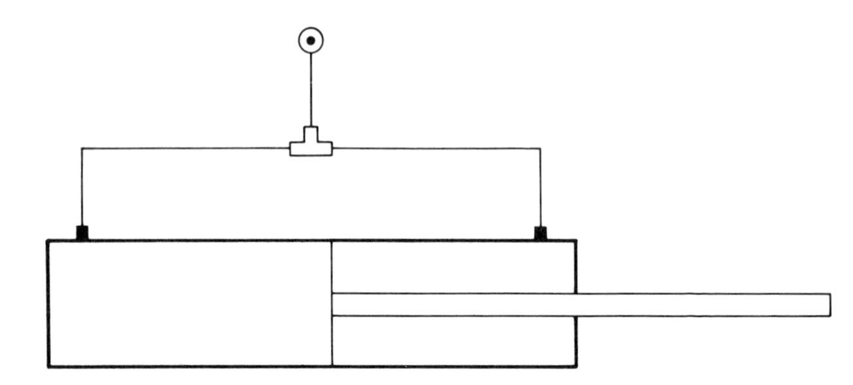

Connect the air supply to both X and Y, but include an on-off valve and a pressure-reducing valve.

2 Place the piston at the centre of its travel.

3 Adjust the air supply to any pressure, and switch on the on-off valve. ***Note what happens.***

4 Repeat 3 for different piston positions and air pressures.

5 ***What conclusions do you draw from your experiment?***

HINT: Two things may help you explain what happens in this experiment.

i) The gradients of the two graphs that you drew in *assignment 3.5*.

ii) A careful study of a pneumatic cylinder, the diagram shown above and the sectional view of a double-acting cylinder shown in the introductory section to pneumatic linear motion.

3.7 Pneumatic Cylinders

To investigate the relationship between the diameter of the piston of a pneumatic cylinder and the force produced

1 Set up the apparatus using a cylinder of approximately 12 mm diameter.

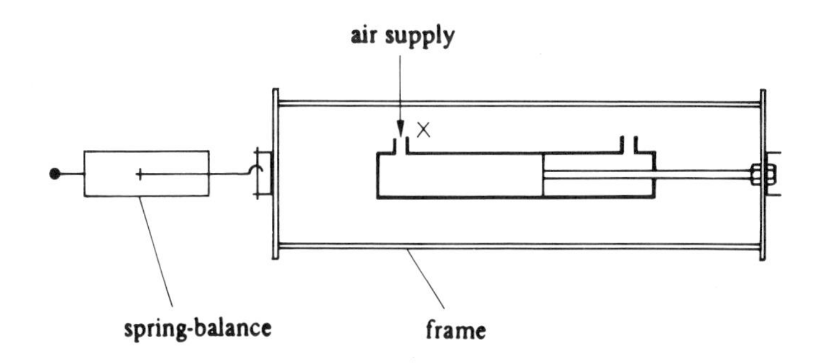

2 Adjust the position of the spring-balance so that it reads zero with the piston rod fully in.

3 Connect the air supply to inlet X, and include an on-off valve and a pressure-reducing valve in the line.

4 Adjust the pressure by means of the pressure-reducing valve to about 3 bar (0.3 N/mm^2) and ***note the cylinder diameter and the spring-balance reading.***

5 Repeat the experiment with say a 20 mm and a 25 mm diameter cylinder, using the same air pressure as before. ***Note the spring-balance reading and the cylinder diameter each time. Make a table for your results.***

6 ***What conclusions do you draw from your experiment?***

NOTE: At the end of your written report, make notes on anything which you consider is important and which you have not mentioned elsewhere.

Pneumatic Control 1

During the linear-motion programmes, you carried out several investigations to discover the operating principles of pneumatic cylinders. We shall now consider the various ways of controlling the air flow into the cylinders by means of pneumatic valves. These valves work in a similar way to the electrical switches which you used to control your vehicle motor.

Pneumatic-control circuit diagrams use standard symbols. You must become familiar with these symbols, and it is suggested that you make a copy of each one in the back of your notebook, for reference purposes.

The three-port valve used in the *Linear-motion assignments* can be actuated in a number of different ways. Your teacher will show you examples of each.

Push button	Suitable for impulse (applied by hand).
Lock-down lever	Hand operated. Remains ON or OFF.
Plunger	Suitable for sensing movement on a machine.
Roller lever	Suitable for sensing movement on a machine.
One-way trip lever	Suitable for sensing movement on a machine (operated by movement in one direction only).
Pilot pressure	Operated by high-pressure air supply (3 bar or greater).
Diaphragm	Operated by a low-pressure air supply (0.2 bar or greater).
Solenoid	Operates when the appropriate electrical supply is supplied to the solenoid coil.

All these valves are returned by a spring. The symbols for these valve arrangements are shown below.

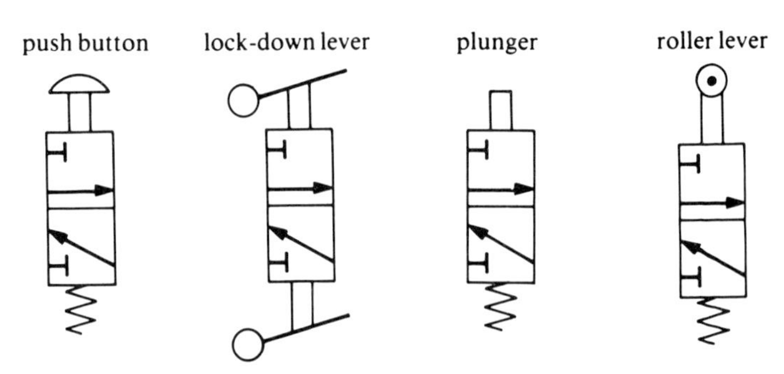

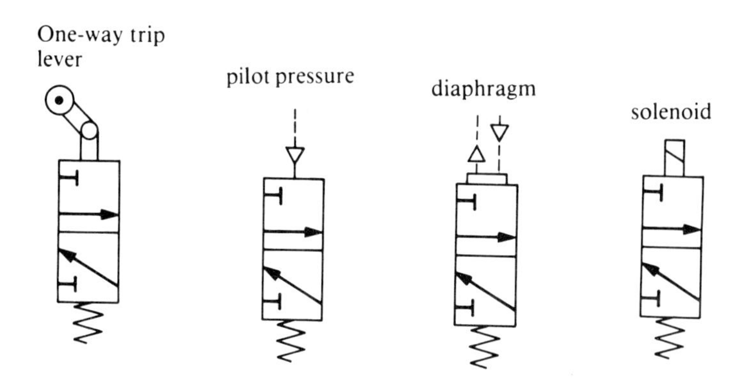

Sketch these symbols in the back of your notebook.

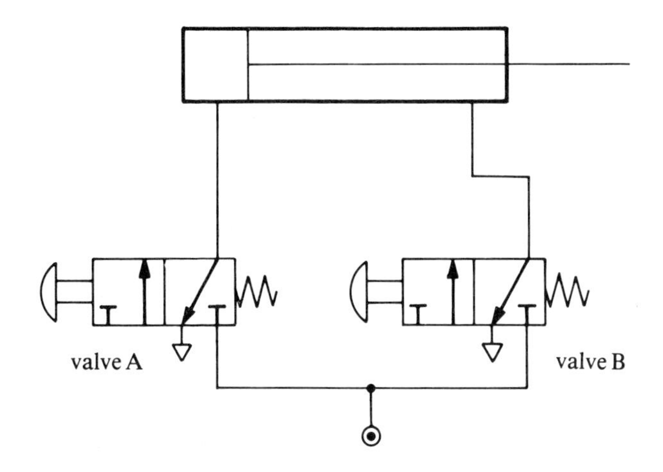

Using any type of hand-operated valve, connect up the circuit shown above to operate a double-acting pneumatic cylinder.

Complete the following table in your notebook

Begin with the piston rod approximately in its central position. Operate the valves as indicated and complete the table in the numbered sequence.

Valve operation	*Movement of piston rod (in or out)*	*Speed of movement (fast or slow)*	*Force produced (small or large)*
1 Valve A off, valve B off			
2 Valve A on, valve B off			
3 Valve A off, valve B on			
4 Valve B on, then valve A on			
5 Valve A on, then valve B on			

2 a) Connect up two lock-down on-off valves to a cylinder as shown below.
Explain why this circuit is unsatisfactory.

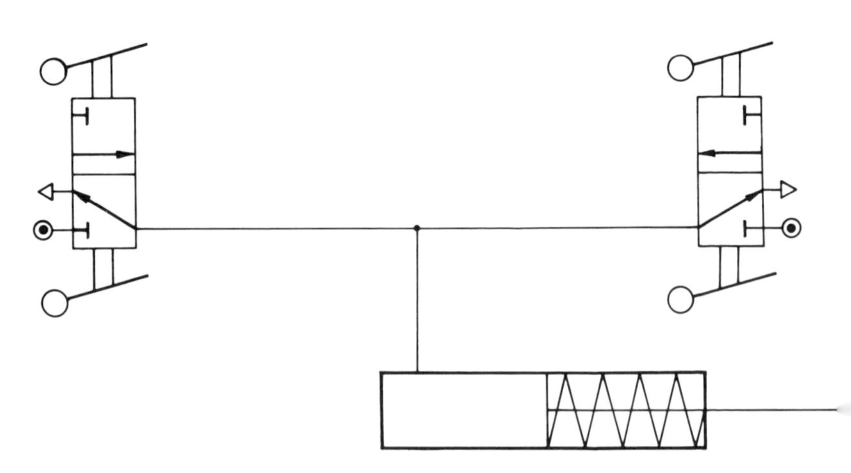

b) i) Investigate the properties of a shuttle valve by connecting one valve and cylinder as shown below.

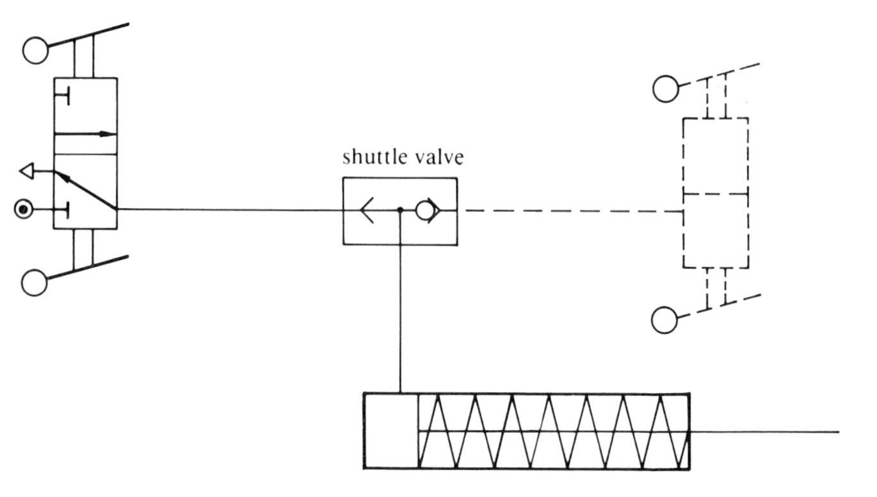

ii) Remove the valve and connect it to the other port of the shuttle valve.

iii) Connect a valve to each of the shuttle-valve ports; you now have a similar circuit to that of (a). Is this satisfactory?

iv) Can you suggest uses for this valve?

Study your results and write a conclusion to this investigation.

Examine the five-port double change-over valve.

The symbol for the body of the valve is:

First position

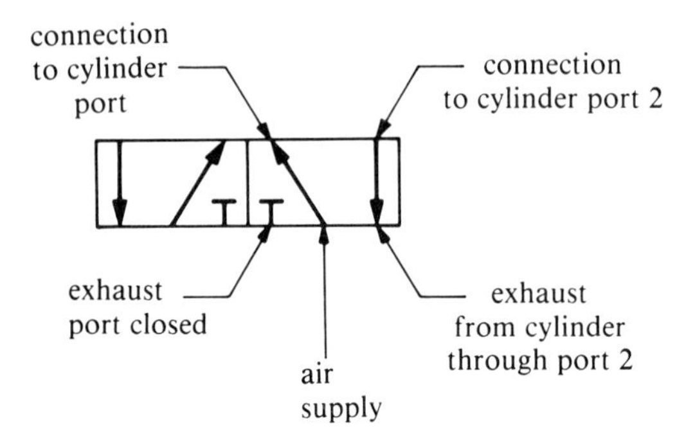

Second position

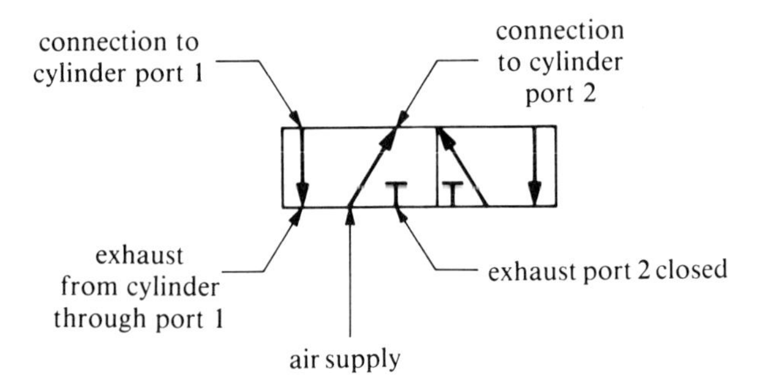

When the valve is in the first position air flows through to cylinder port 2 while cylinder port 1 is open to exhaust and the exhaust port 2 is closed. When the valve is changed to the second position air flows through to cylinder port 1 and it is cylinder port 2 that is open to exhaust and exhaust port 2 that is closed.

This type of valve may also be actuated in a number of ways, for example:

push-button	one-way trip lever
lock-down	pilot (by an air signal)
plunger	spring (for return only)
push-pull	diaphragm (low pressure air signal)
roller lever	solenoid

By adding the appropriate actuator to the valve body, the symbol for the valve is obtained.

Example

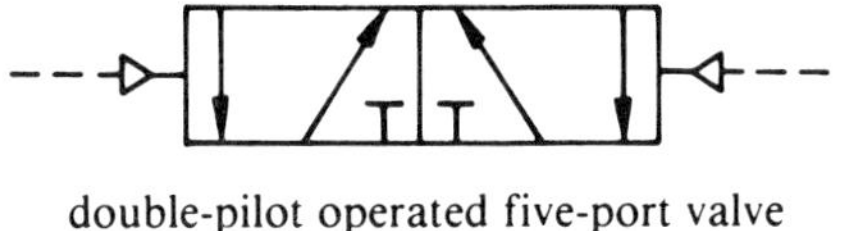

double-pilot operated five-port valve

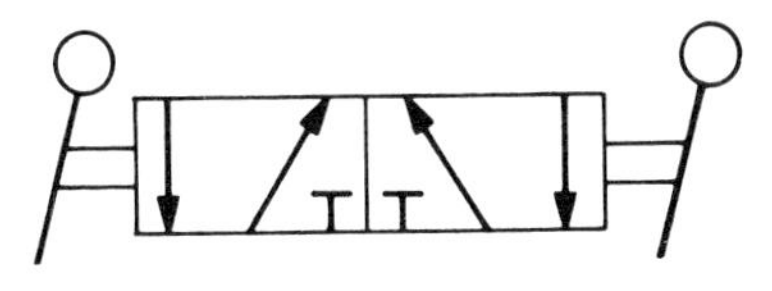

lock-down lever operated five-port valve

ir-pressure actuators, for example

single-pilot, spring return, and

double pilot,

e most commonly used on this type of valve, and the symbols for ese are shown on page 58.

Single-pilot spring-return double change-over valve

Air pressure (greater than 3 bar) applied at pilot port A will cause the valve connections to change over, when the air pressure is removed, the valve returns to the position shown.

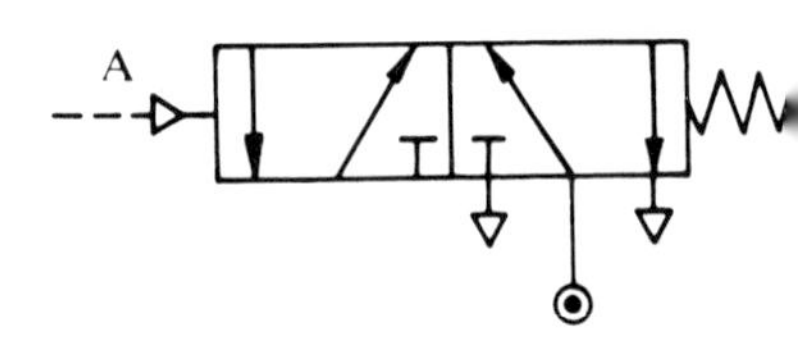

Double pilot-operated double change-over valve

Air pressure (greater than 2 bar) applied at pilot port A will cause the valve connections to change over. The valve will remain in this condition even when the air pressure is no longer applied at port A.

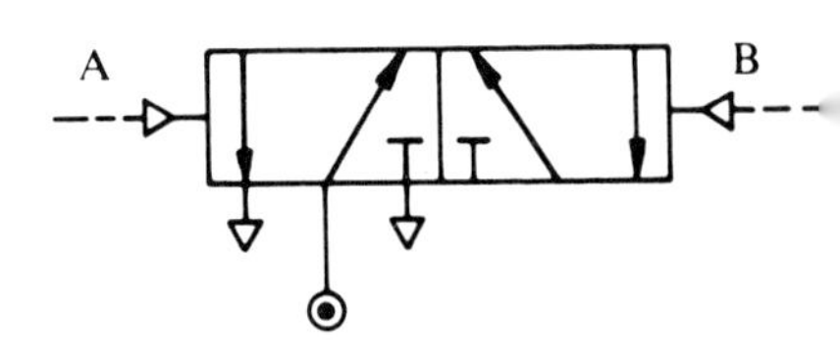

Air pressure at pilot port B will return the valve to its original position, provided that the pressure has been released at pilot port A.

What kind of relay circuit is similar in operation to a double pilot-operated double change-over valve?

1 Connect up the circuit shown below, using a hand-operated five-port change-over valve to operate a double-acting cylinder

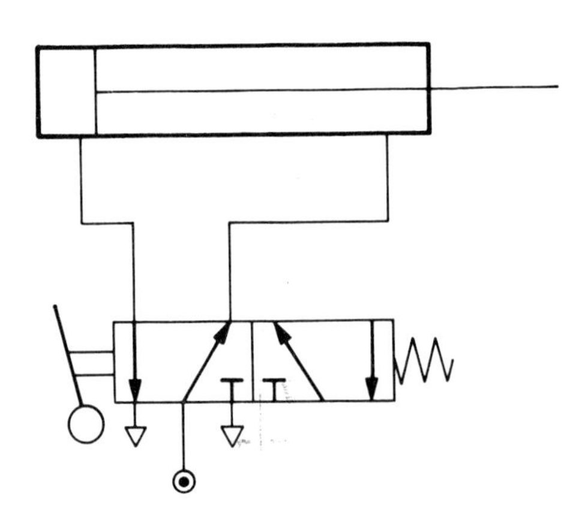

2 Using two hand-operated on-off valves and a double pilot-operated double change-over valve, ***design in your notebook*** a circuit which will make the piston rod go out when an impulse from one on-off valve is applied and which will make the piston rod return when an impulse from the other on-off valve is applied.

NOTE: Circuit diagram lines indicating pilot operating connections are drawn as broken lines.

Connect up your circuit and try it.

3 Bolt the cylinder on to a piece of plywood, or mount it on a 'Hybridex' frame, and carefully position and fix two valves (plunger or roller-actuated) so that one is actuated as the piston rod reaches the end of its outward movement and the other is actuated when the piston rod is retracted.

Example

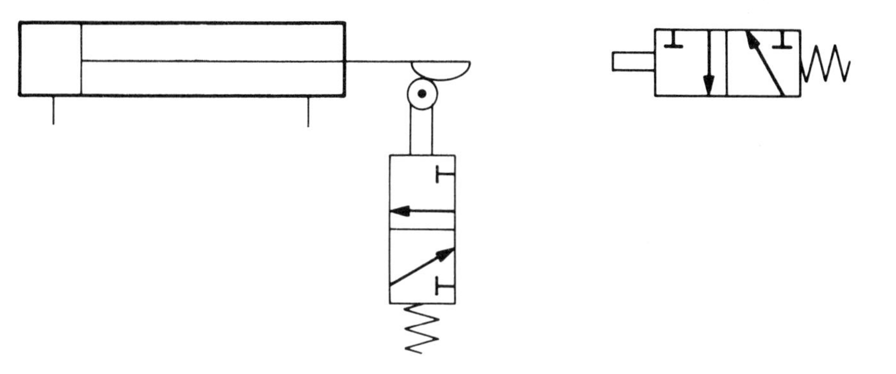

ow, using the circuit you designed in 2, you should find that the iston rod oscillates in and out continuously while the air supply is onnected.

With the circuit used in 3, the only method of stopping the piston rod from oscillating is to switch off the air supply, which is unsatisfactory since the piston rod:

i) cannot be stopped in a known position,

ii) can be easily pushed into any position when the cylinder is not pressurised.

Study circuit 3 and, by introducing two lock-down on-off valves modify it so that the piston rod can be stopped at any time in either the in or out position *with the cylinder pressurised*.

The circuit should still operate automatically if both lock-down valves are on.

Also, if one valve is placed in the on position and the other is used to supply a short impulse, only one cycle of the operation should occur.

When you have succeeded, ***draw the circuit in your notebook.***

Pneumatic Control 3

You are now able to control the direction of movement of the piston-rod of a double-acting cylinder, but it is also often necessary to control the speed of movement of the rod.

For example, when drilling a hole, the feed movement must be fairly slow to prevent the drill from being broken, yet the return motion should be fast to save time.

In *Pneumatic control 1* you discovered that, by applying pressure on both sides of the piston, the speed of movement of the piston-rod was slowed down considerably (and at the same time the force produced was greatly reduced).

A convenient method of controlling the speed of movement of the piston-rod of a double-acting cylinder is by restricting the flow of air from the exhaust side of the piston. This is done by means of a *variable restrictor*.

Symbol

Record this for future reference.

In practice, the restrictor is fitted in the appropriate exhaust port on the control valve, as shown in the diagram below.

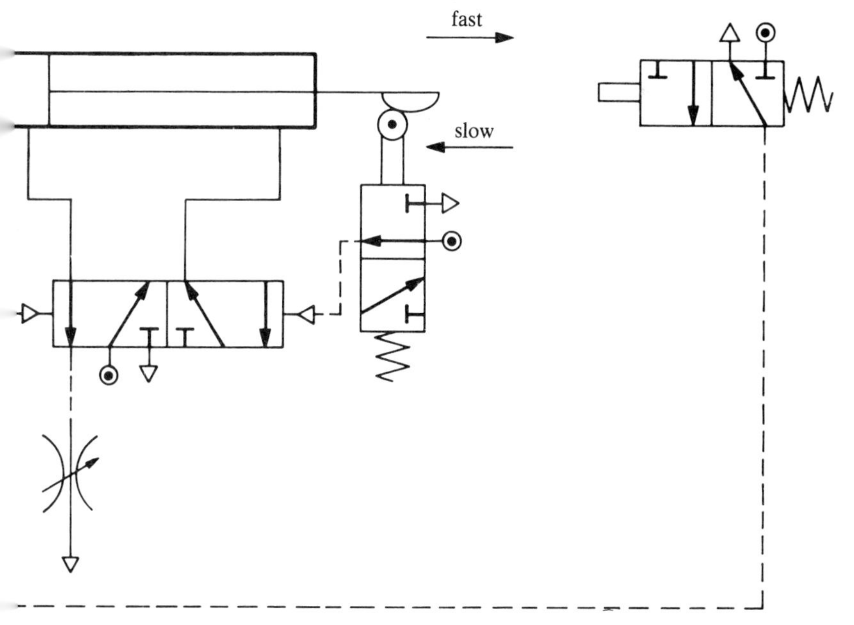

The speed of movement is controlled by adjusting the air flow through the restrictor.

In addition, a second restrictor can be fitted into the other exhaust port, to control the speed of movement of the piston-rod in the opposite direction.

1 Connect up the oscillating circuit shown above fitting an exhaust restrictor into each exhaust port. Adjust the restrictor settings to give variation of piston-rod speed in either direction. ***Draw the complete circuit in your notebook, and explain its operation.***

For the double-acting cylinder the exhaust restrictor provides an easy and cheap means of controlling piston-rod speed. Can the same method be used for a single-acting cylinder? Think about this.

The *flow-control valve* is again a variable restrictor, but this fits into the air supply line and limits the flow of air *in one direction only*. In the opposite direction, full-flow conditions are maintained The direction of full and restricted flow is clearly marked on the valve.

Symbol

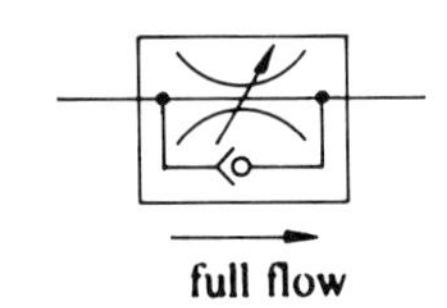

Record this for future reference.

2 Connect a flow-control valve between a single-acting cylinder and an on-off valve. Adjust the flow-control valve so that the piston-rod moves out slowly when the pressure is applied. ***Draw the arrangement in your notebook, giving it a suitable title.***

3 Remove one exhaust restrictor from the circuit used in 1 and, by inserting a flow-control valve, modify the circuit so as to give the same results (e.g. speed control of the piston-rod in either direction).
Draw this circuit in your notebook.

A three-port, single change-over valve is also available with the same range of actuators.

Symbol

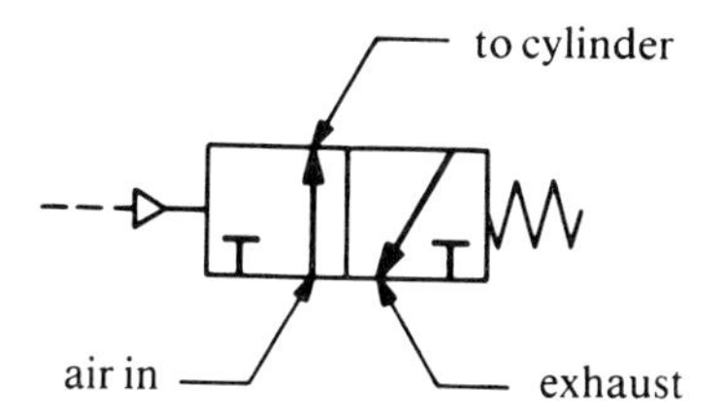

Example: single pilot spring-return three-port spool valve

Record this in the back of your notebook, together with the other symbols.

Normally the cylinder port is open to exhaust and, when operated, the inlet is connected to the cylinder as with the three-port on-off valve. However, with the spool type of valve the function of the ports may be interchanged — the air supply may be connected to either the inlet or the exhaust port.

If the air supply is connected to the exhaust port, the cylinder will be pressurised with the valve unoperated and exhausted with the valve operated — the inversion of the normal function.

Normal

Apply air pressure at X to operate the cylinder.

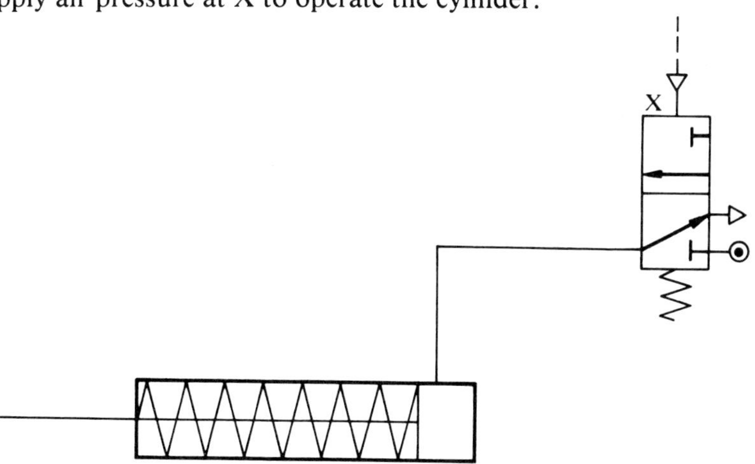

Inversion

Apply air pressure at X to retract the piston-rod.

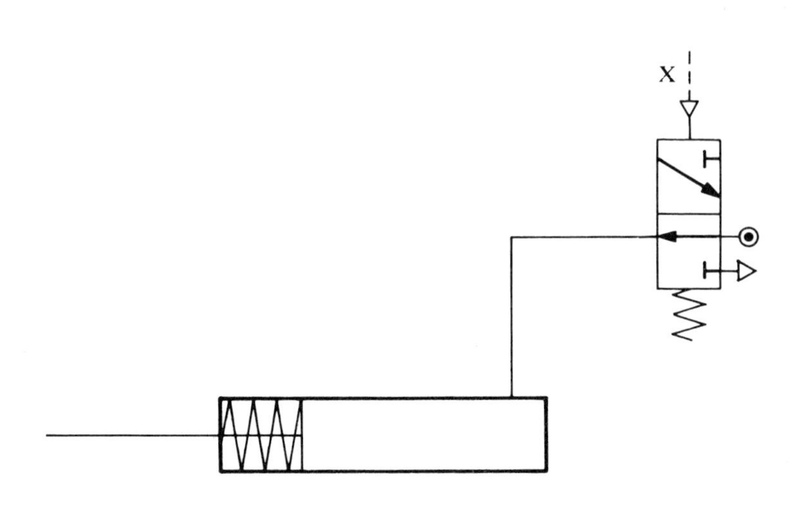

Note that the exhaust and inlet ports have been reversed.

Electrical equivalents

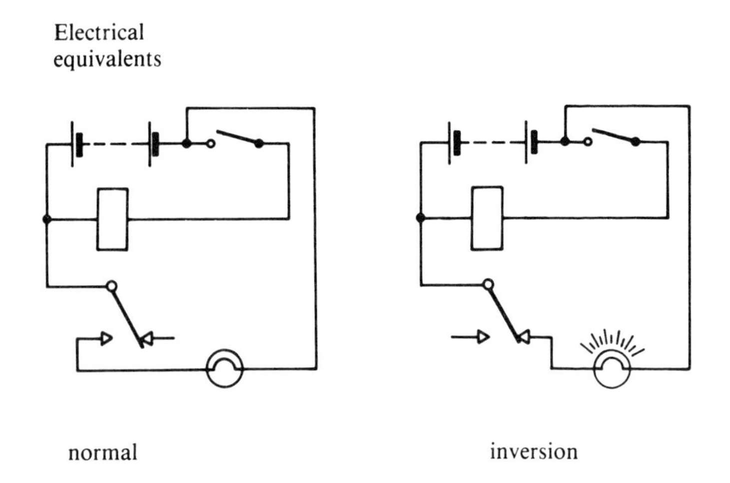

The output from the valve (cylinder port) need not necessarily be to a cylinder. In a circuit shown below, a reservoir (a container which stores a volume of compressed air) is connected to the cylinder port of a three-port single change-over valve.

Symbol — for a reservoir

Record this for reference purposes.

1 ***Copy the following circuit in your notebook.***

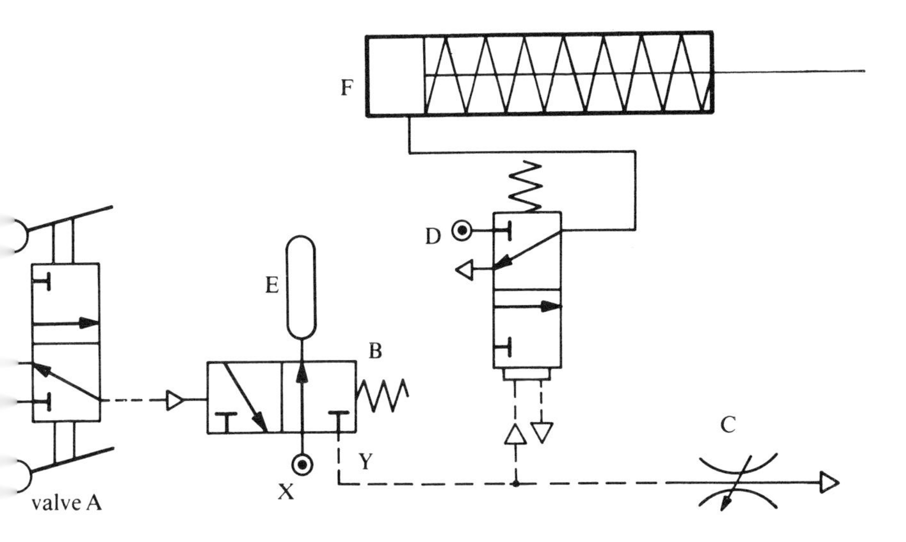

Connect up the circuit and:

- Switch ON valve A for a short time, i.e. apply an impulse. ***Note what happens, and try to explain the operation of the circuit.***
-) Switch ON valve A and leave it ON. ***Again note what happens, and explain why.***

iii) Alter the exhaust-restrictor setting and repeat (i) and (ii). Does the circuit give exactly the same results as before?
If not, describe how and why it differs.

iv) *Would the fitting of a larger reservoir affect the circuit?*

2 Now interchange the pipe connections X and Y of valve B and *draw the modified circuit in your notebook.*

   i) Apply an impulse at valve A.

   ii) Switch ON valve A for a longer period of time (several seconds), then switch it OFF.

1 Investigate a further circuit which includes a reservoir and a flow-control valve.

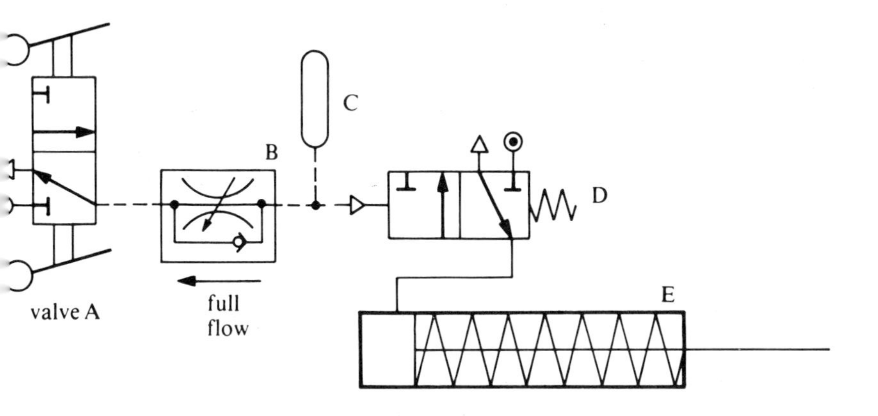

Copy the circuit into your notebook.

Connect up the circuit and:

i) Apply an impulse at valve A.

ii) Switch ON valve A for a longer period of time, then switch it OFF. In each case, observe carefully what happens. ***Record your observations, and explain the operation of the circuit.***

iii) Adjust the flow-control valve and repeat (i) and (ii). If the results now differ from those obtained before, ***describe the effect and offer an explanation.***

iv) ***Would the fitting of a larger reservoir affect the circuit?***

v) ***Would the replacement of valve D by a diaphragm-actuated valve affect the output of the circuit?***

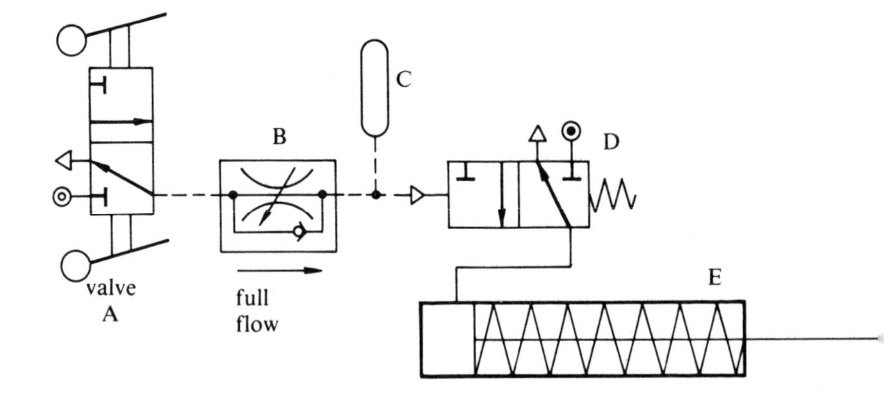

2 ***Draw the circuit in your notebook with the flow-control valve reversed as shown above.***

i) Apply an impulse at valve A.

ii) Switch ON valve A, and leave it ON.

iii) Make adjustment to the flow-control valve and repeat. ***Explain the effect this has on the circuit.***

iv) ***Would the fitting of a larger reservoir affect the circuit?***

Pneumatic Control 6

The following circuit, which could be used to control a drilling machine, uses a combination of the simple circuits with which you are now familiar.

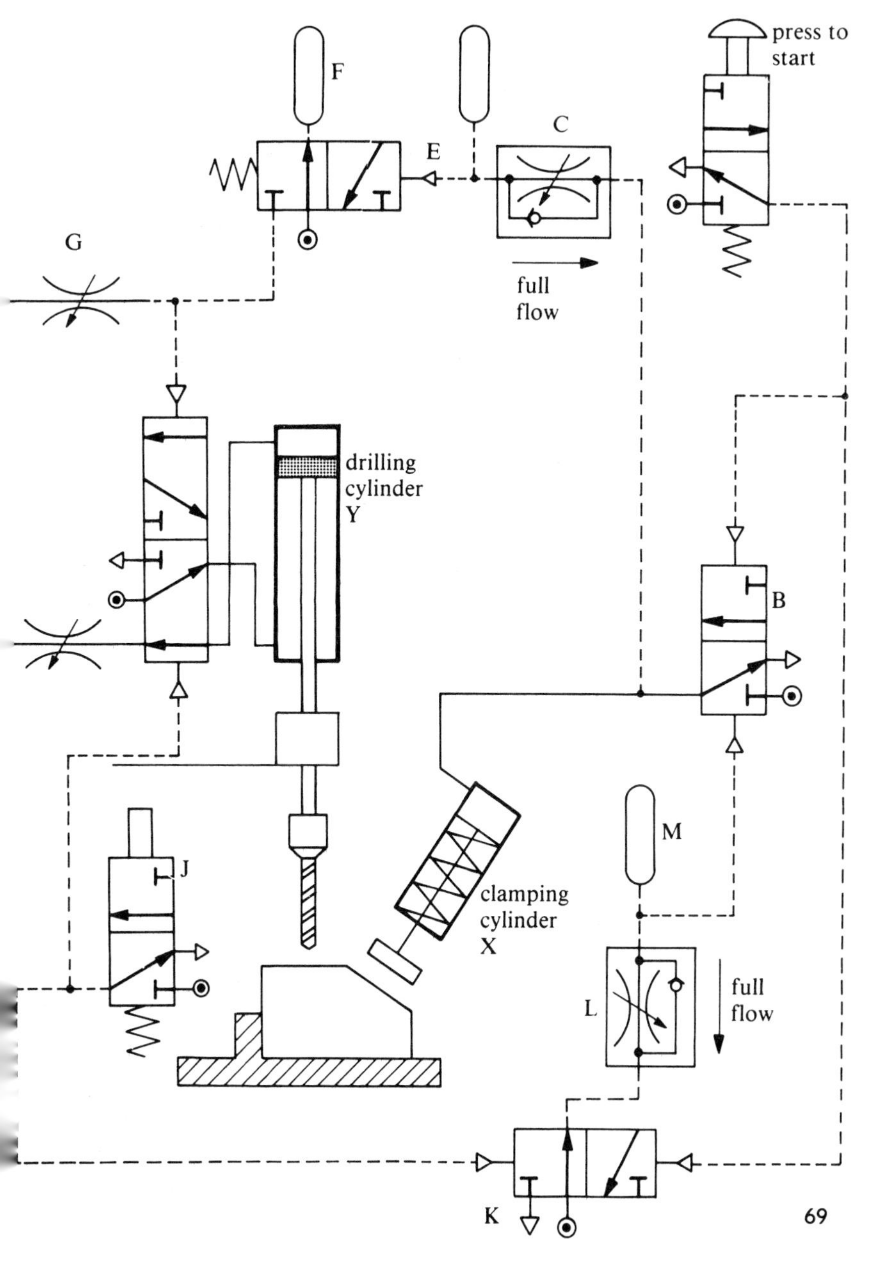

1 Study the circuit thoroughly, ***then list, in the correct sequence, the movements of the clamping and drilling cylinder.***

2 ***Explain as fully as you can how the circuit operates.***

3 If sufficient components are available, connect up the circuit and check that it operates as you expected.

4 Did you have difficulty in making the circuit operate? ***Can you suggest any disadvantage of this circuit?***

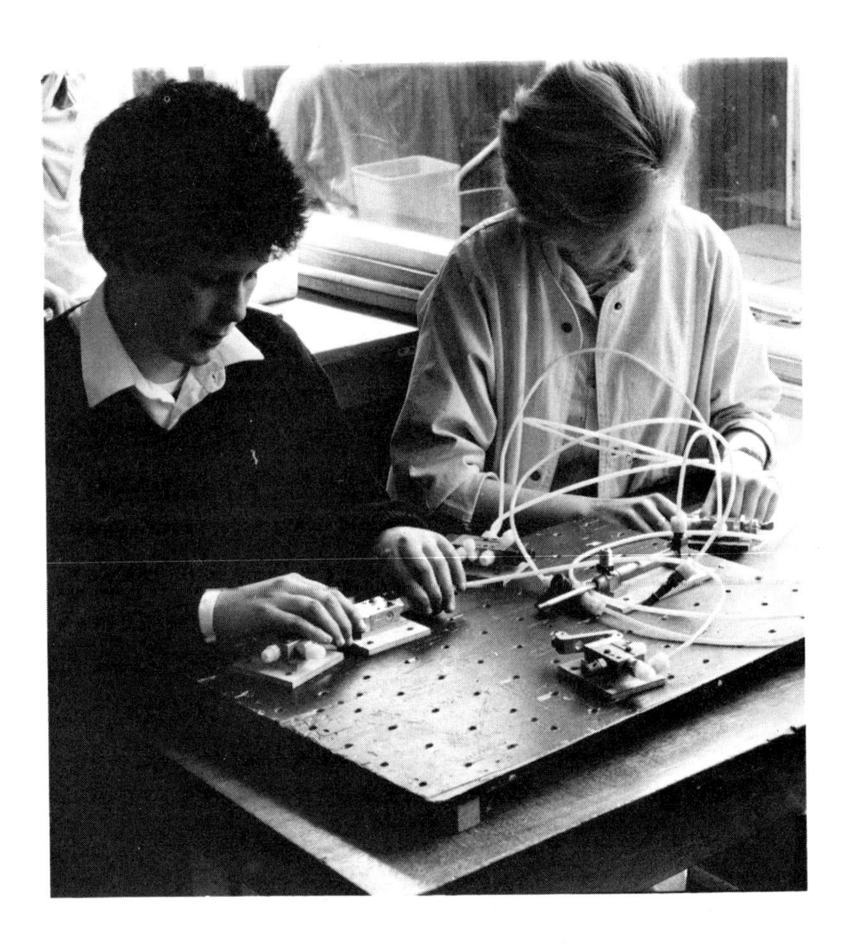

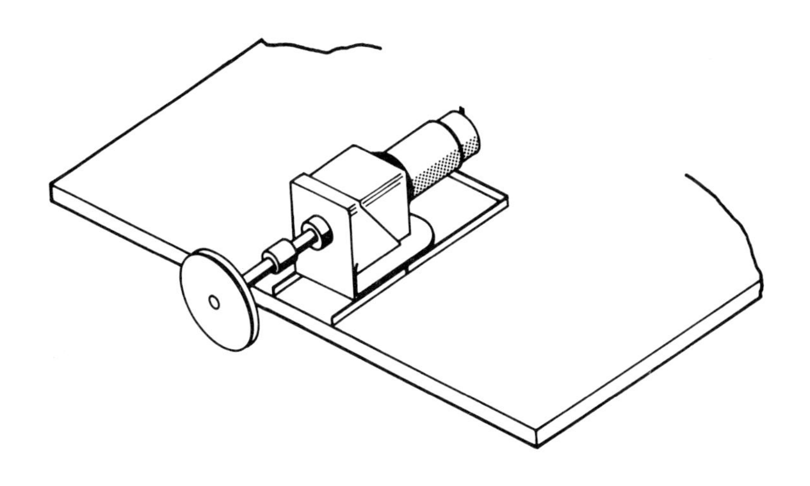

1 You have been given an electric motor complete with a 360:1 ratio gearbox. Fit a Meccano wheel to the shaft and clamp the motor to the bench. On occasions you may have found that, although the motor is fitted with a gearbox, you cannot adjust the speed of the output shaft to an exact value. In the investigations you are about to undertake, you will discover ways of making the speed of your motor continuously variable, and you will find explanations for the methods used.

2 Connect the motor to a nominal 12 volt d.c. power supply, and estimate the speed of the wheel (We use the term 'nominal' because the output from power supplies varies under certain conditions.) Using suitable meters (0–20 V and 0–1 A d.c. ranges should be suitable), find the voltage across the motor terminals and the current flowing in the motor. (The voltage across a pair of terminals is often called a 'potential difference'.)

Record in your notebook the values you have obtained, using a table as shown below. With this arrangement, the values in the 'resistance' column A and the potential difference across the resistor, column C, will be zero.

A	B	C	D
resistor value (ohms)	*potential difference across motor* (volts)	*potential difference across resistor* (volts)	*current in circuit* (amperes)

3 Fix a box, containing a 25 ohm resistor, in series with one of the motor leads and switch on the supply.

Note the effect on the speed of the motor. Does it run faster or slower, or does it continue running at its normal speed?

4 Place a voltmeter across the motor terminals and measure the voltage now supplied to the motor. Measure the current flowing in the circuit.

Record the values in your notebook table.

5 Measure the voltage across the resistor (potential difference) and ***note the value.*** What explanation can you now give for your observations in 3?

6 ***What is the relationship, if any, between the p.d. across the resistor, the p.d. across the motor, and the supply voltage?*** (A circuit diagram showing the voltages you have measured will help you here.)

7 If the p.d. across any resistance in any circuit is divided by the current flowing in the resistance in the circuit, the value obtained is called the 'resistance' of the resistor and is measured in *ohms* (symbol Ω — the Greek letter omega).

Divide the p.d. across your resistor labelled 25 ohms by the current flowing in it, by referring to your table, and you should obtain a value somewhere in the region of 25 Ω.

The resistor will be marked with a 'tolerance' value, e.g. 5%. Determine whether or not the value you have calculated is within ± 5% of 25 Ω. Plus or minus (±) 5% means the actual value may be 5% above or below the value given.

Record your findings.

8 Now work out the effective resistance of your motor when it is connected in series with a 25 Ω resistor, and ***record the result.***

9 Place two 25 ohm resistors in series with the motor, and observe the effect on the motor speed.

Remove the 25 ohm resistors and replace them with a single 50 ohm resistor, and note the motor speed. How do the two speeds compare?

From your observations, what would you say is the total resistance when two 25 ohm resistors are connected in series?

Record any conclusions.

10 Connect a voltmeter across the extreme ends of the two series resistors, measure the current in the circuit, and calculate the resistance of the two resistors in series. How does the calculated value compare with your conclusions in the last part of 9?

Record your calculations and comments.

11 You have measured the resistance of a resistor using an ammeter and voltmeter. Is it possible that this resistance could change under certain conditions? Think about this.

1 You have learned that the resistance of an electrical component can be found by setting up a suitable circuit and measuring the current flowing in the resistance and the potential difference (voltage) across the resistance. You were asked to think about whether or not resistance could change under certain conditions. Did you have any ideas? If you are able to, set up some apparatus to test your suggestions.

2 Take a light source which uses a 12 volt bulb, and connect it to a 1 volt supply. Measure the voltage across the bulb and the current flowing in it. Work out the resistance of the tungsten filament. Increase the voltage to about 12 volts, take both readings, and again work out the resistance. How do the two resistances compare?

In your notebook, record what you have done and try to come to some conclusions from the resistance values you have obtained.

3 There are two ways of connecting a variable resistance into a circuit: the rheostat connection and the potential divider (sometimes called the potentiometer connection).

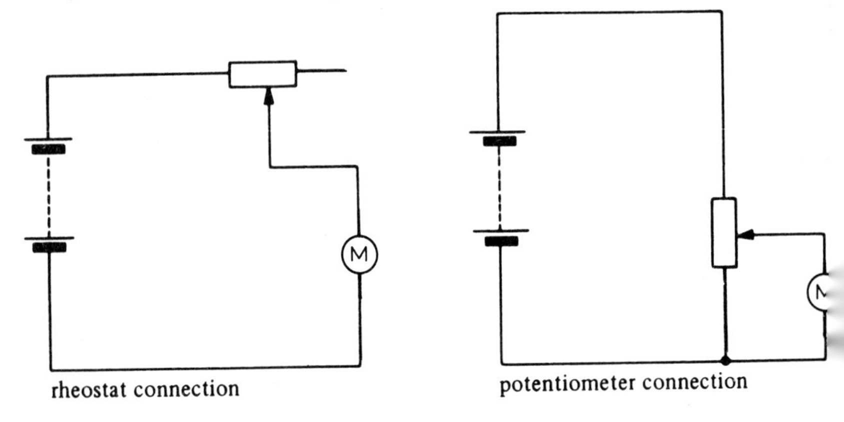

rheostat connection

potentiometer connection

Fit a 100 ohm rheostat in series with the motor on the turntable. Vary the resistance of the rheostat and note the effect. Fit a voltmeter across the motor and an ammeter in the circuit and again vary the rheostat.

Record your observations. Try to account for what you observe.

4 Adjust the variable resistor to make your motor rotate at approximately half a revolution per second. Remove the voltmeter from across the motor terminals and place it across the rheostat. Calculate the resistance needed to make your motor rotate at this speed.

Record the result.

5 An ohmmeter is often used to measure resistance. Find the setting for the rheostat which allows the turntable to run at half a revolution per second. Switch off, and now measure the resistance of the rheostat at this setting, using an ohmmeter. Compare the reading with the value you have calculated in 4.

An ohmmeter is basically an ammeter movement in series with a rheostat and a battery (fitted internally in the instrument).

Procedure

To use the ohmmeter, first short out the meter terminals and adjust the rheostat until the resistance reads zero on the ohms range. This is called 'zeroing' the ohmmeter, since a short-circuit across the meter terminals represents zero resistance. The setting of the rheostat is then at the correct value to allow a current just sufficient to cause the pointer to go to 'full-scale deflection'. Any additional resistance now placed in the circuit will give a smaller current. Resistance values are marked off on the ohms range. (Note that this is a non-linear scale.)

To measure a resistance, place it across the ohmmeter terminals, after zeroing, and read off the resistance from the scale. If the meter terminals are left open-circuited, there is no circuit and the pointer does not move. In this position the ohmmeter is marked ∞ (infinite resistance).

Ohmmeters vary considerably in construction, and you should consult your teacher if you are unsure about the correct setting.

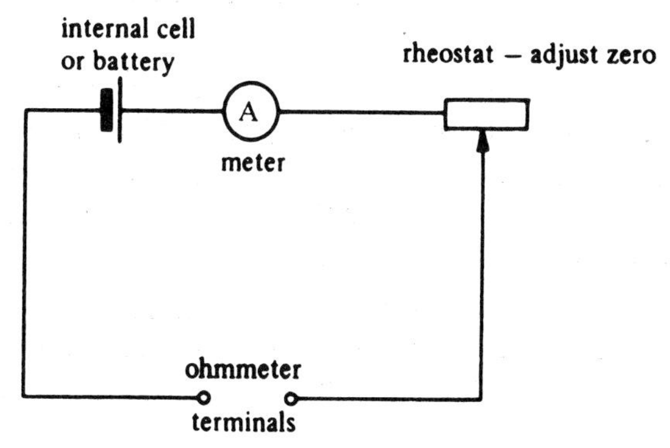

1 Measure the actual output voltage (terminal potential difference) of a nominal 12 volt d.c. supply when the supply is connected only to the voltmeter. ***Note the reading.*** We will call this reading the 'off-load' voltage.

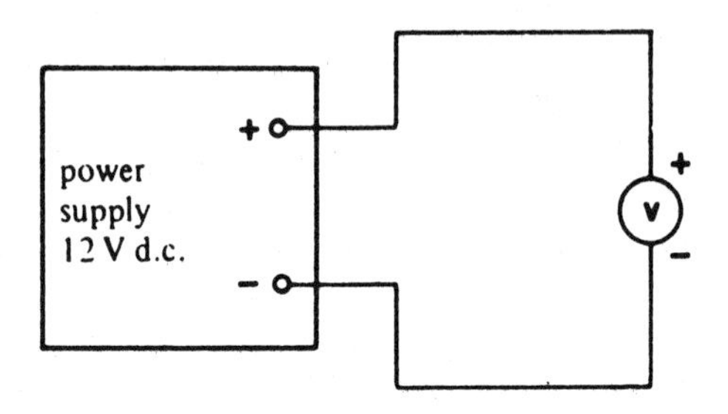

2 Leaving the voltmeter in circuit, connect a 12 volt solenoid across the supply terminals.
Note the reading on the voltmeter. This is the 'on-load' reading, the load being the current consumed by the solenoid.

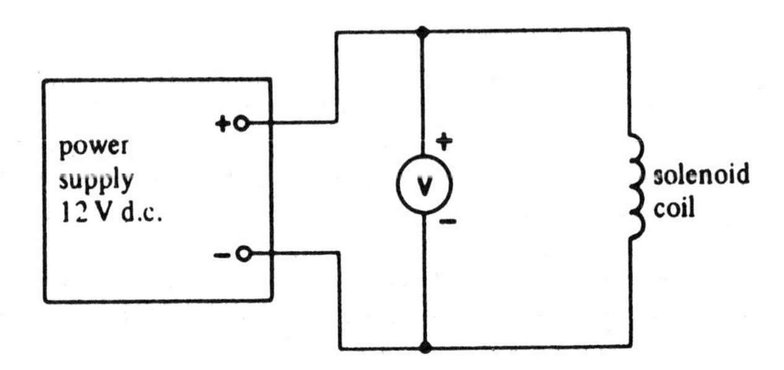

3 Remove the solenoid from the circuit and replace it with a 25 ohm resistor. ***Note the terminal potential difference.***
Replace the 25 ohm resistor with a 50 ohm, a 75 ohm, and a 100 ohm resistor in turn. ***In each case, observe and note the effect on the terminal p.d.***

4 Measure the 'off-load' voltage at the terminals of a 12 volt lead acid accumulator, and compare it with the 'on-load' voltage when the battery is connected to the solenoid you used in 2.

Warning. These accumulators can deliver very high currents. Take great care not to short circuit the accumulator because the wires used would immediately become red hot. Accumulators contain a strong liquid acid which you must be careful not to spill.

5 What can you say about the terminal p.d. of some power supplies, and what explanation can you give for the effects you have observed in assignments 1 to 4?

6 Place a 25 ohm resistor across the nominal 12 volt d.c. power supply. Place a finger on the resistor (inside the box) as soon as you have switched on.

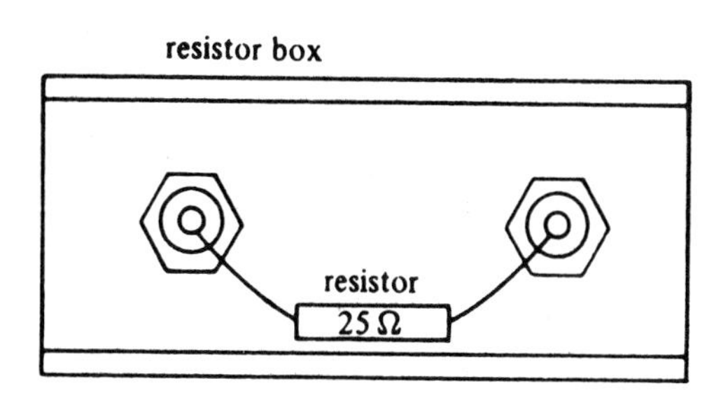

7 Repeat 6 using a 50 ohm resistor.

8 The 25 and 50 ohm resistors are 'wire-wound', i.e. they consist of a length of thin wire, which has a high resistance compared with copper wire, wound round a 'former' of insulation material (usually ceramic — made from clay). Connect a 'carbon' resistor of about 10 ohms resistance across the supply, and observe the effect. There is no need to put a finger on the resistor this time! Now use a 10 000 ohm carbon resistor.

Make a note of the effects you have seen in the test in 6 to 8, and try to draw some conclusions. List them 1, 2, 3, etc.

1 When using carbon resistors, you will have found that usually the value is not stamped on them in figures. Instead, you will have noticed that coloured bands are used. These bands indicate the value according to the 'resistor colour code'.

Resistor colour code

1st and 2nd digit code		*multiplier code*
gold	not used	× 0.1
black	0	× 1
brown	1	× 10
red	2	× 100
orange	3	× 1000
yellow	4	× 10 000
green	5	× 100 000
blue	6	× 1 000 000
violent	7	
grey	8	
white	9	

tolerance code	
brown	1%
red	2%
gold	5%
silver	10%
no colour	20%

A salmon-pink band indicates that the resistor is of the high stability type, i.e. the value indicated in code will not exceed the tolerance value under normal circuit conditions.

The physical size of a resistor gives a guide to its power rating, though this is not always the case since new manufacturing techniques allow the production of higher wattage components with smaller physical size.

Note that gold is used as a multiplier or as a tolerance value. The position of the gold band indicates its specific use. It is used as a multiplier for values less than 10 Ω.

2 The diagram shows the method of colour coding such resistors.

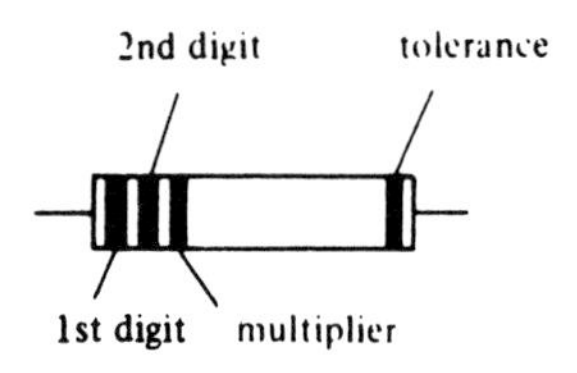

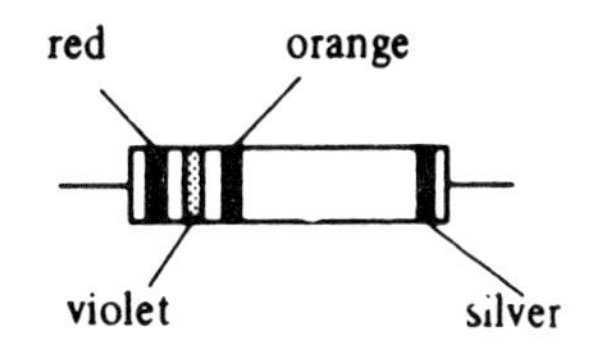

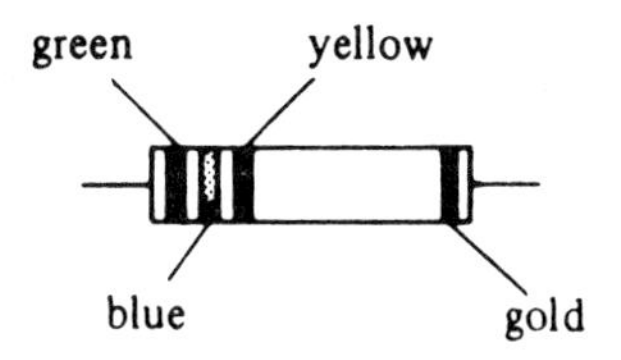

i) 1st digit: red — 2
2nd digit: violet — 7
Multiplier: orange — 3
Tolerance: silver — 10%

Resistance value is
27×1000 ohm $\pm$ 10%
i.e. 27 000 ohm $\pm$ 10%

ii) 1st digit: green — 5
2nd digit: blue — 6
Multiplier: yellow — 4
Tolerance: gold — 5%

Resistance value is
$56 \times 10\,000$ ohm $\pm$ 5%
i.e. 560 000 ohm $\pm$ 5%

The terms 'kilohm' and 'megohm' are commonly used to indicate 'thousands' and 'millions' of ohms.

e.g. 27 000 Ω can be written 27 kΩ
10 000 000 Ω can be written 10 MΩ

3 You have been given ten coded resistors. Identify the values in ohms and the tolerances by making use of the table shown.
Make a copy of the table in your notebook.

1st digit colour	*value*	*2nd digit colour*	*value*	*3rd digit colour*	*value*	*tolerance colour*	*resistance* (ohm)	*tolerance* %

4 Using a 50 ohm wire-wound resistor box, find the resistance of the resistor as indicated by an ohmmeter.
Note the value obtained (the value should be within 5% of 50 ohm if the ohmmeter is accurate and if the resistor is correctly marked).

5 Now fit a second 50 ohm resistor in parallel with the first, and find the effective resistance of two 50 ohm resistors connected in parallel.
Note the value obtained.

6 Repeat 4 and 5 using 25 ohm resistors. If other values are available, use them, but ensure that both resistors tested are of the same value.

What can you say about the effective value of two similar resistors connected in parallel?

7 Connect a 25 ohm and a 50 ohm resistor in parallel, and measure the effective resistance.

Now try connecting any pair of resistors in parallel, and measure the effective value of each combination.

What can you say about the effective value of two dissimilar resistors connected in parallel?

Introduction: Using the cathode-ray oscilloscope

When you have measured voltages in the past, you will have used a voltmeter, usually of the moving-coil type. These meters are not sensitive to changes in voltage if the voltage is changing at a fast rate. The cathode-ray oscilloscope enables you to detect any rapid change in the voltage.

1 You will have been shown an oscilloscope cathode-ray tube. Small negative charges called electrons are emitted from a heated *cathode* and are attracted towards the screen by applying a positive voltage to an *anode* (final anode) further down the neck of the tube. As the electrons accelerate towards the final anode, they pass a focusing anode which, by adjustment of the voltage on it, enables the stream of electrons to be focused to a fine spot on the screen. The screen is coated with *phosphors* which give out light when bombarded with fast-moving electrons. The colour of the light depends on the kind of phosphor used, typical colours being white (television tube), blue, amber, and green.

Your teacher will have switched on an oscilloscope, showing a stationary spot on the screen, or he will probably tell you how to obtain a spot on your own. Notice that you can not only *focus* the spot by varying the voltage on the focusing anode, but you can also vary the brightness by making the *grid* more positive than the cathode, thus increasing the number of

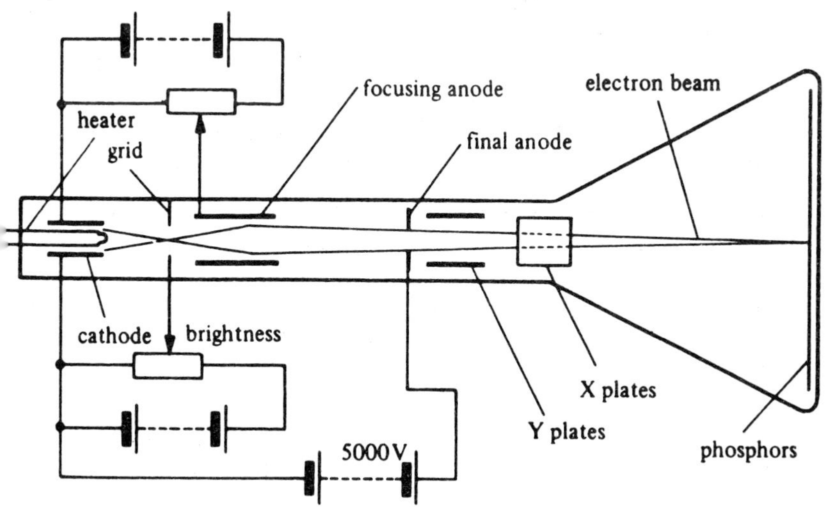

electrons in the stream. Never leave a very bright stationary spot on the screen, since damage to the phosphors may result.

2 Circuits in the oscilloscope enable the spot to be moved horizontally and vertically. If you examine a typical oscilloscope cathode-ray tube, you will probably see two pairs of plates at about the point where the diameter of the tube increases. The rear pair of plates are the 'Y' plates, to give vertical deflections, and the pair nearer the face of the tube are the 'X' plates, to give horizontal deflections. The diagrams show how the spot is shifted if voltages are applied to the X and Y plates.

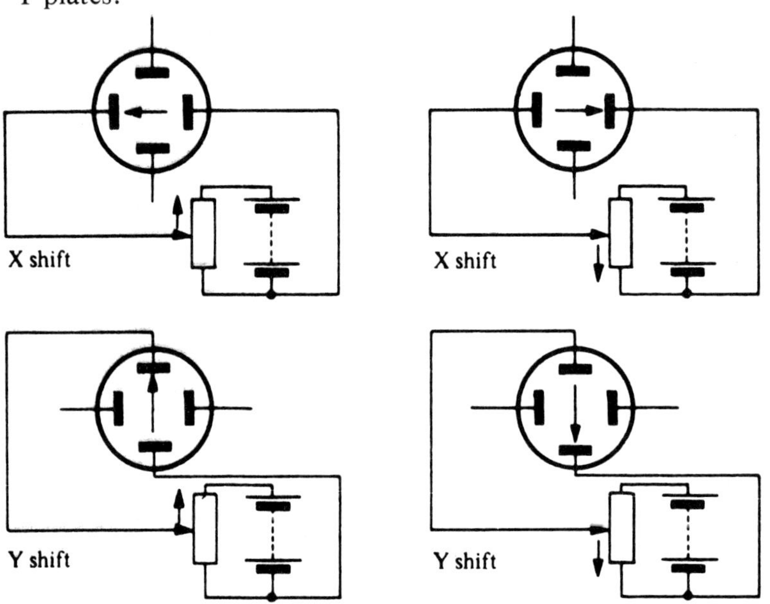

Locate the positions of the X- and Y-shift controls. Adjust them, and note the effect on the position of the spot.

3 The X plates are also connected to internal circuits which will drive the spot across the face of the tube at a uniform velocity (i.e. if the spot moves half-way across the face of the tube in 1 second, it will take 1 second to cover the other half). The spot returns very quickly to the start when it has completed the scan. The scanning velocity is variable, and can be varied in two ways:

i) by switch for coarse variations, i.e. large steps; and
ii) by a variable resistor which acts as a fine control.

Locate the coarse velocity-control switch and rotate it through all its positions, noting the effect on the velocity of the spot. Now adjust the fine velocity control in all the positions of the coarse switch.

4 The Y plates are not only connected to the Y-shift controls but also to an amplifier, if required. If amplification is not required, then the Y plates are coupled directly to an input socket.

Locate the Y-input socket and the amplifier-input socket. These sockets are used to feed in the voltages which are to be examined. Always try the Y-input socket first and, if the picture is too small vertically, use the amplifier input and adjust the *gain* control until you have a suitable *trace*.

5 Adjust your oscilloscope to obtain a spot centrally on the screen. Connect a 6 to 12 volt d.c. supply (preferably a battery) between the Y-input terminal and the common 'earth' terminal. Note the effect on the spot position. Reverse the supply.

NOTE: Some oscilloscopes are fitted with an a.c./d.c. switch, and for this test the switch must be in the d.c. position. The diagrams show the effect of the battery voltage.

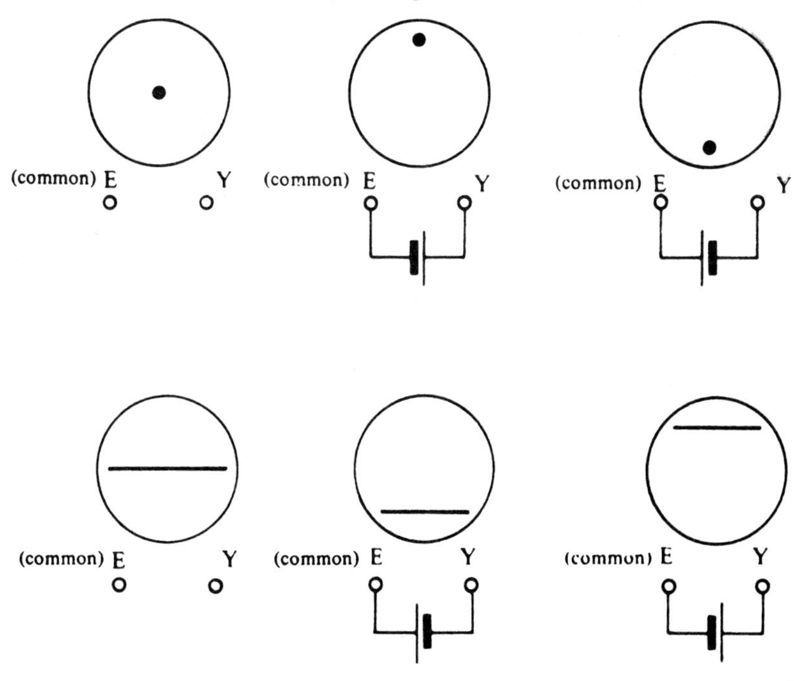

6 Leaving the battery connected, make the spot move across the screen, using the velocity controls. Reverse the battery polarity.

The diagram shows the sort of trace you should obtain. Since this trace is not in the centre, a voltage must be present; and since the line is straight, the voltage must be constant at all times. The X trace is often called a 'time base', since it shows how the voltage on the Y plates is varying with time.

7 In the assignments to follow, you will be using the oscilloscope to examine voltages which are continuously changing with time.

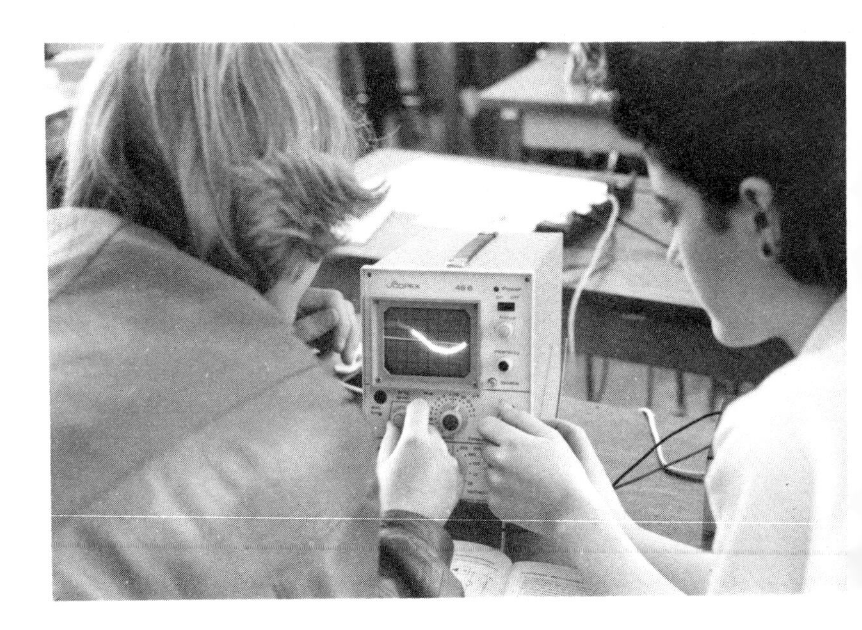

Electronics: Rectification 1

1 You will have been given a box containing a transformer. This device works on a.c. (alternating current) only, and if you were to supply it with high voltage d.c. it could burn out. The function of a transformer is to 'step-up' or 'step-down' the voltage applied to it. Your transformer is a step-down type, since it steps down 240 volts from the mains to 12 volts. The 240 volts is fed to the *primary* coil and the 12 volts is taken out from the *secondary* coil.

2 Connect your electric motor to a 12 volt d.c. supply, and ensure that the motor functions normally. Now connect the motor to the 12 volt a.c. outputs of your transformer. Note that there are two separate 12 volt outputs, either of which can be used. Make these connections briefly since your motor may be damaged if you have it connected for more than a few seconds.

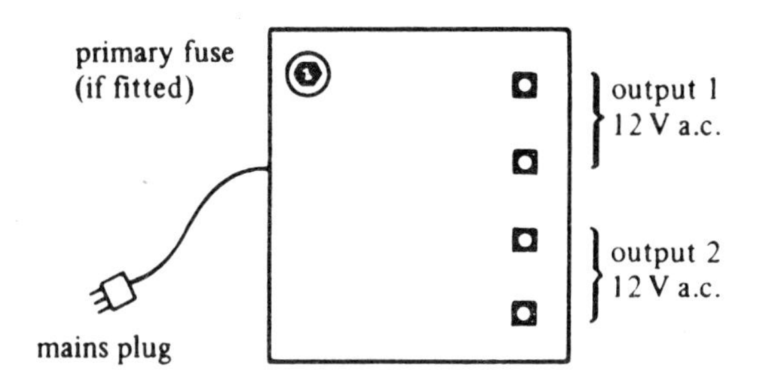

Note what happens to the motor

Feed the 12 volt d.c. supply into your oscilloscope and observe the trace — the motor should not be connected.

Now observe the trace of 12 volt a.c. supply on the oscilloscope.

Draw the traces you obtain, and try to explain the behaviour of the motor in 2.

4 Connect a 'diode-box' in series with your 12 volt a.c. supply and the motor, as shown.

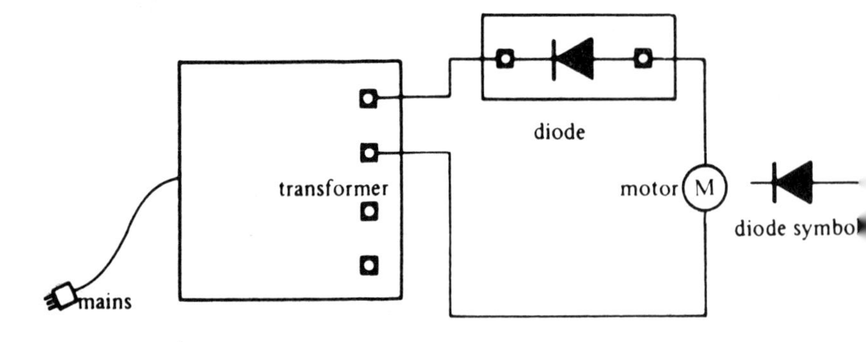

Switch on and observe what happens.

Reverse the diode and observe the effect on the motor performance.

5 Repeat what you have just done in 4, but this time use the oscilloscope to observe the voltage being fed to the motor.

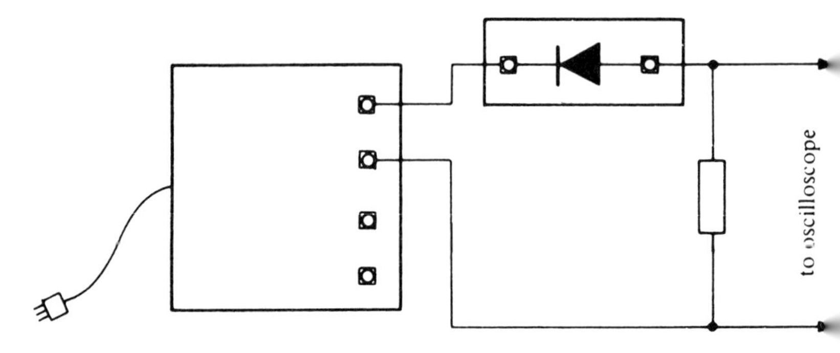

Draw your oscilloscope traces.

What effect is the diode having on the a.c. supply? Use a voltmeter to test your ideas if you wish. The oscilloscope traces you have previously observed should help you to understand th a.c. trace and also to understand the function of the diode in your circuit.

In addition, to help you to understand the function of the diode, devise a 'reversing' circuit for the diode so that by quickly operating a manual switch you will be able to observe the effect on the alternating current.

6 The oscilloscope traces you have just seen are not very clear, since the motor brushes do not make perfect contact and produce 'noise' on the trace. You will already have discovered in a previous programme that the resistance of the motor when it is running is in the order of 50 ohms. Replace the motor with a 50 ohm, 3 W resistor to act as a load, and repeat 5.

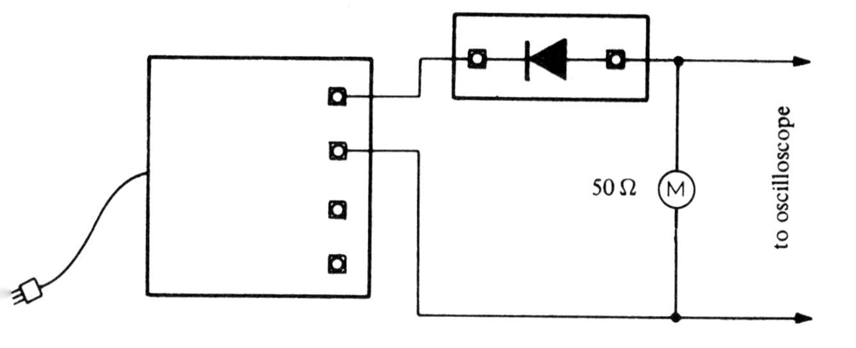

Electronics: Rectification 2

1 You have found that d.c. can be produced from a.c. by making use of one rectifier diode. Half-wave rectification is produced, since the d.c. output consists of only one half of the input waveform, the other half being lost. Two other methods of rectification are commonly used, each offering advantages and disadvantages:
   i) full-wave rectification,
   ii) full-wave 'bridge' rectification.

2 Set up a full-wave rectification circuit as shown below.

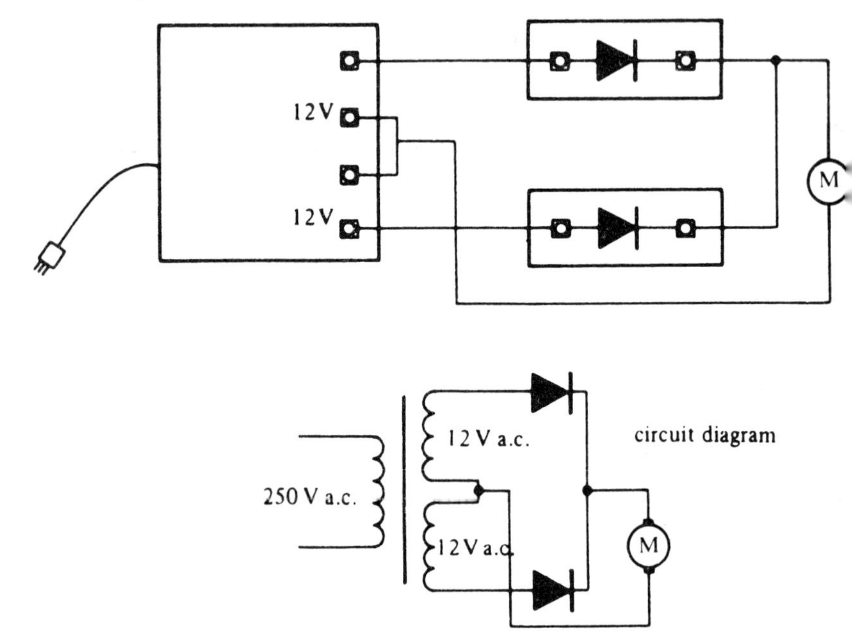

Observe the output-voltage waveform across the motor on your oscilloscope. Replace the motor with a 50 ohm resistor unit to act as a load (approximately equal to the motor resistance), to eliminate the effect of 'brush noise'. Again observe the output waveform. Reverse the diodes.

Draw in your notebook the waveforms you have seen, together with a description of how you modified the circuit to obtain each waveform. What would you say is the purpose of the othe diode in the circuit, compared with the single diode in half-wav rectification?

3 Repeat 2 using a full-wave 'bridge' circuit as shown below.

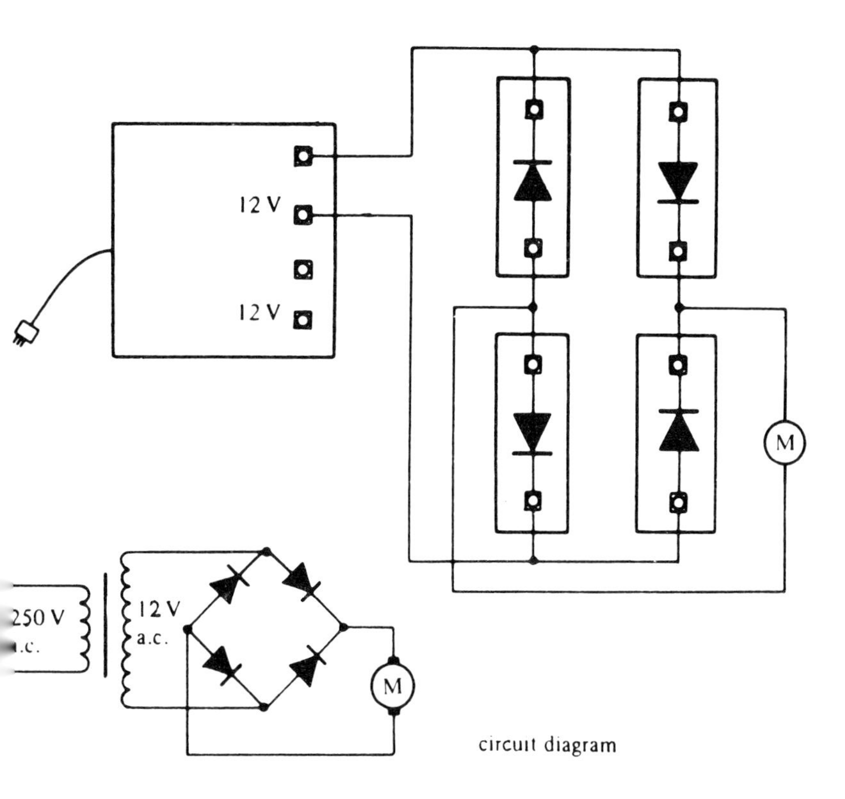

circuit diagram

How do the full-wave and full-wave 'bridge' output waveforms compare?

What are the two major differences in circuitry between the full-wave and full-wave 'bridge' circuit?

How do the motor speeds compare when using half-wave, full-wave, and full-wave 'bridge' rectification? If you have not made a note of this, set up each circuit again. It is not necessary to measure the speed accurately each time, but if there is a noticeable difference then make a note of it.

Replace the motor with a 50 Ω resistor and observe the output waveform from the mains-transformer secondaries as shown below. Arrange a switching circuit using a switch-box which will enable you to switch the Y input on the oscilloscope (CRO) to either the top or the bottom of the transformer secondary fairly rapidly.

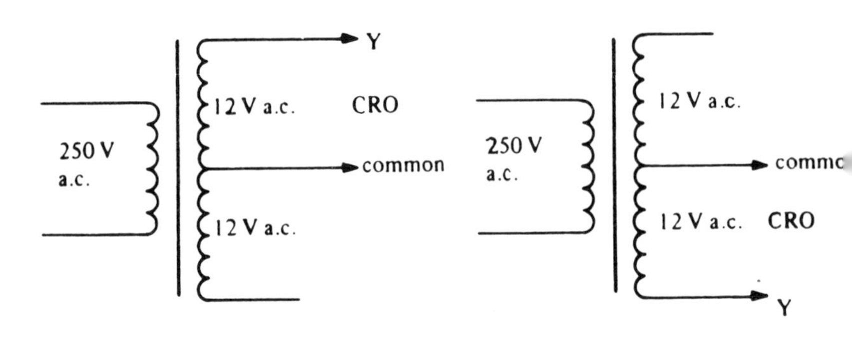

Draw the waveforms you see, and try to work out how a full-wave rectifier functions in terms of 'conventional' current flow (By convention we can assume that electrical currents flow from positive to negative.)

7 Connect up a rectifier circuit — any one of three will do — and measure the current taken by the wheel motor.

Now 'load' the motor by slowing the wheel a little with your finger — but not excessively — and ***note the effect on the current consumed from the rectifier circuit.***

You have found that, when the motor is loaded, the current required by the motor increases, and consequently the current delivered by the rectifier circuit must increase. Remember also that the current supplied by a power supply can be adjusted by fitting a suitable value of load resistance across the supply terminals.

Set up a half-wave rectifier circuit using the motor as a load. Fit a 5000 μF (approximately) capacitor across the d.c. output terminals and ***note the effect on the output voltage and the motor speed. Note: Observe the polarity of the capacitor!***

Using your oscilloscope, observe the voltage waveform being applied to the motor, both with and without the capacitor.

Replace the motor with a 50 ohm 5 W resistor and again observe the output waveform.

Draw the oscilloscope traces in your notebook.

Set up a full-wave and a 'bridge' rectifier circuit in turn, observing the effect on the output voltage, the motor speed, and the oscilloscope trace in each case.

In your notebook, describe what you have seen, and draw the appropriate oscilloscope traces.

Try to give explanations for any changes in output voltage, motor speed, and oscilloscope trace when a capacitor is fitted

It is often desirable that the output voltage of a power supply should remain fairly constant over a range of load currents. If the output voltage of one circuit remains more constant than another over a range of load currents, then we say that this circuit has a better 'regulation'. Using different values of 5 W load resistors, devise experiments to find which of the three common types of rectifier circuit has the best regulation and which the worst. Use a 'smoothing capacitor' of about 5000 μF for your tests.

In your notebook, give an account of your tests and findings.

Most of the d.c. waveforms you have examined so far have not been of 'pure' d.c. but of 'pulsating' d.c. Many circuits, such as radio and television circuits, require relatively 'ripple-free' d.c.

Set up a half-wave rectifier circuit to drive your motor. Fit a capacitor of about 1000 μF across the supply output terminals, and note the effect on the 'ripple' as the motor is loaded by braking the wheel with a finger. Try other values of capacitor if you wish.

5 Using different values of load resistance and smoothing capacitor, make tests to determine the following.

i) If a power supply is needed which has a low ripple current, must the value of the smoothing capacitor be changed if the supply is to deliver a higher current?

ii) For a particular load current, which of the three common rectifier circuits requires the least value of smoothing capacitor for a fairly smooth (ripple-free) output voltage, and which requires the largest value?

In your notebook, give an account of your tests and findings in each case.

Examine the transistor unit by turning the unit over to expose the transistor. Each of the three connections to the transistor is attached to a 4 mm socket, as follows:

i) the emitter — connected to the black socket,
ii) the base — connected to the green socket,
iii) the collector — connected to the red socket

The transistor in the unit is a power type and, in common with other power transistors, the collector lead is internally in contact with a metal plate to which a metal or plastic cover is fitted. The base and emitter leads are insulated both from the collector and from each other. Note also that the body of the transistor (collector) is bolted directly to an aluminium plate.

What do you think is the purpose of this plate? You have seen a similar plate used in the diode unit.

Ask your teacher to show you other types of transistors.

The symbol for the type of transistor we are using is shown here alongside the diagram of the unit.

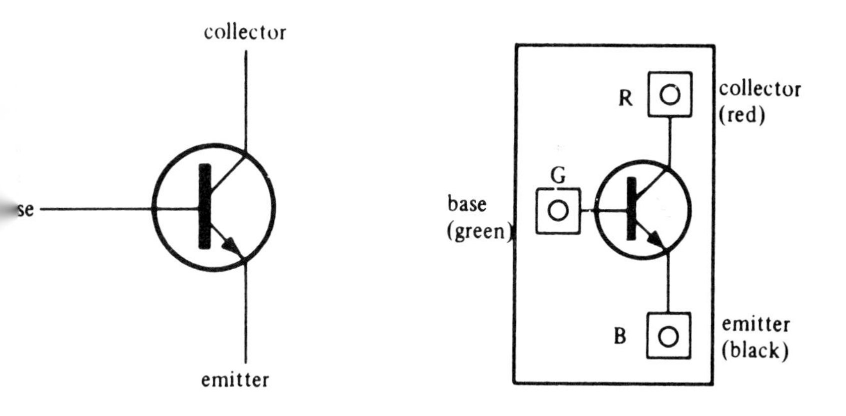

Using an ohmmeter, measure the approximate resistance between each of the three sockets, a pair at a time. After each reading, reverse the ohmmeter leads and take a second reading.

Make a suitable table for your results, and record them in your notebook. It is sufficient to quote the readings as 'low' — a few tens of ohms, 'fairly low' — a few hundred ohms, or 'high' — a few thousand ohms.

The readings you have taken are for a 'good' transistor. Should you at any time suspect that your transistor is damaged, test it and compare the results with those in your table.

4 Take a diode unit and test the diode using the same procedure as for the transistor. (There are only two sockets this time.)
What similarities or differences are there between the resistances of diodes and the resistances of transistors?

5 Set up the circuit (circuit A) shown in the diagram below, such that the emitter (black socket) is wired to the negative terminal of the voltage supply and the collector (red socket) is wired through the 12 volt lamp to the positive terminal.
NOTE: The transistor may be damaged if the voltage supply connections to the collector-emitter circuit are interchanged. Remember this in all your work using transistors!

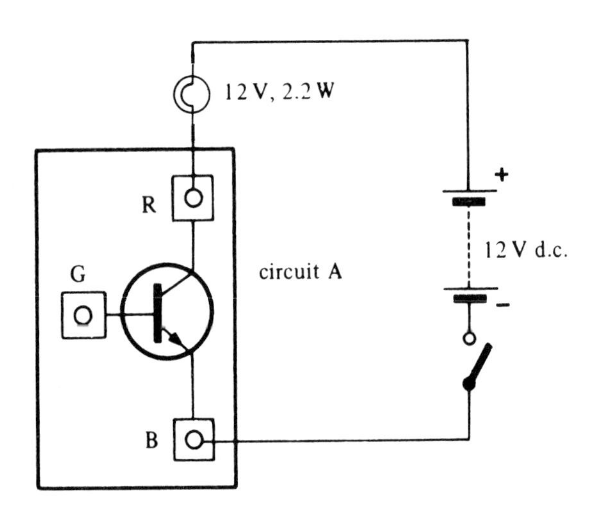

Switch on the supply. *Note what happens to the lamp.*
Is sufficient current flowing in the circuit to light the lamp?
Do your observations agree with your transistor-resistance measurements?

6 Now add a second circuit (circuit B) between the base and emitter of the transistor, as shown. The second supply should be a 1.5 volt cell.

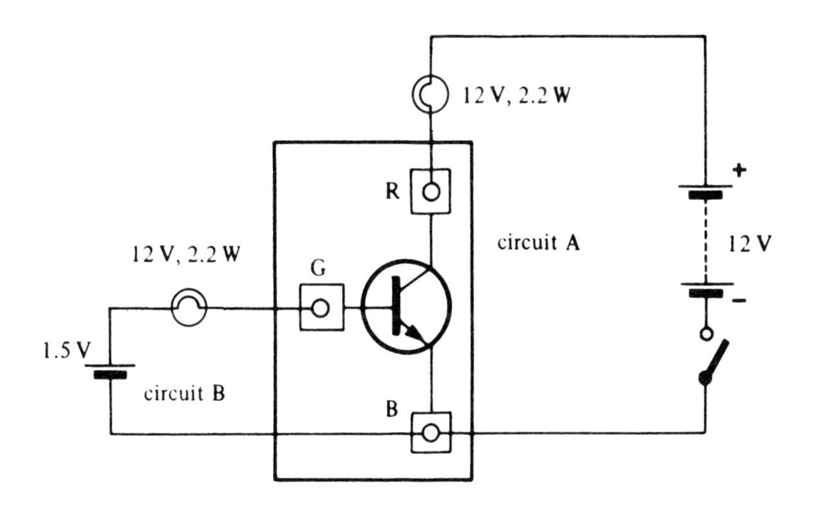

Switch on the supplies and note what happens to both lamps.

Reverse the 1.5 volt cell in circuit B — *not* the voltage supply in circuit A.

Note what happens.

7 Again reverse the 1.5 volt cell, so that the positive is once more connected to the lamp and the negative is connected to the transistor emitter (black socket).

Fit an ammeter (approximately 0–0.5 A, or use a multimeter) first in the base-emitter circuit and then in the collector-emitter circuit.

Note the ammeter reading in each circuit.
How do you account for the behaviour of circuit B?
Briefly summarise your findings from your transistor investigations to date.

1 Set up the circuit shown below, using a 1.5 volt cell in the base — emitter circuit. Also include a 25 000 ohm (25 kilohm) variable resistor, as shown, in the rheostat connection.

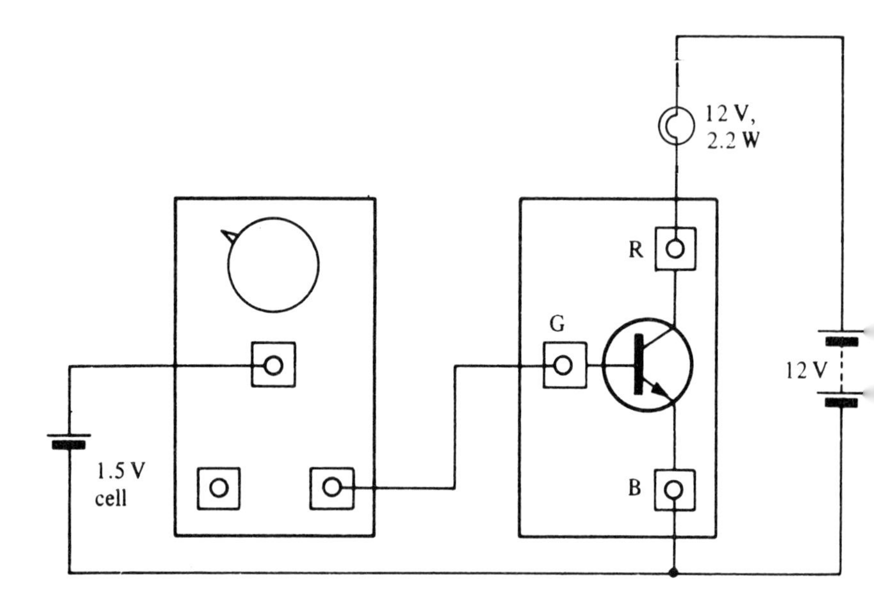

Note the effect as you change the resistance of the rheostat by turning the control knob. Include a circuit diagram in your notebook, in addition to the above block diagram.

2 Fit an ammeter (0–1 A at first) into the base circuit, and observe the effect on the base current as the rheostat is adjusted from end to end.

Remove the ammeter from the base circuit and fit one in the collector circuit (0–1 A). Again adjust the rheostat. (If you have two ammeters available, you need not remove the ammeter in the base circuit unless you wish to do so.)

3 Replace the lamp in the circuit with a motor and vary the base current with the rheostat. Remove any ammeters if you wish.

You have seen previously that a transistor can be used as a switch, rather like a relay. From the tests you have just made, can you suggest other uses for a transistor?

Until now you have made use of two separate power supplies to supply the electrical energy for a transistor. However, one supply can be used to provide both the base and collector currents, as shown here.

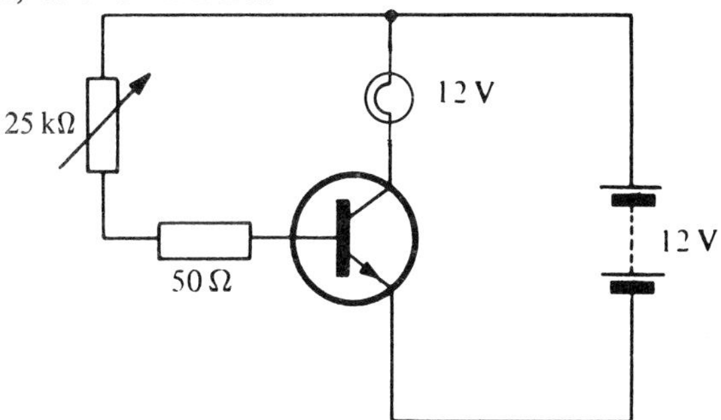

Your previous circuits used a 1.5 volt cell in the base-emitter circuit. This low voltage did not provide a sufficiently high current in the base-emitter resistance to damage the power transistor. The 50 ohm resistor is included in this new circuit in order to limit the base current (base-emitter current) to a safe value when connecting a 12 volt supply. In later programmes you will be setting up circuits of your own design and, unless instructed otherwise, you must *always* include such a resistor with a value of at least 50 Ω.

Set up the circuit and test it. Replace the lamp with a motor and again test.

Record your tests and the results.

Set up a speed-control circuit, using only the motor, a battery, and a 100 ohm variable resistor and no transistor.

Compare the transistor method of speed control with the simpler rheostat method.

Replace the 25 kilohm variable resistor with a photocell. Use the cell as

i) a switch, and
ii) a variable resistor

to drive an electric motor and then a 12 volt lamp. If you wish to vary the brightness of a photocell light source, make use of a 100 ohm variable resistor in the rheostat connection.

In your notebook, draw suitable circuit diagrams and attempt to give full explanations of how they work in terms of base and collector currents.

1 Set up the following circuit:

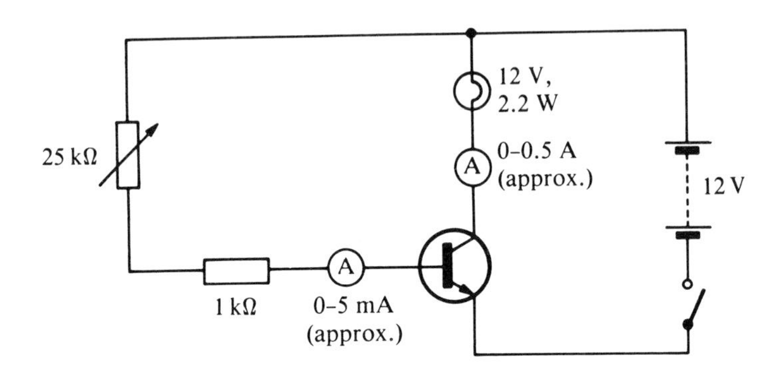

Before switching on, adjust the rheostat to 'maximum' resistance. Switch on, and adjust the rheostat to give satisfactory readings on the two current meters.

NOTE: The 12 volt 2.2 watt lamp acts as a protection device. limits the maximum current that can flow through the device t about 180 mA. For this reason you should not take meter readings in this assignment when the collector current exceeds 150 mA because they can result in incorrect values of h_{FE}.

Note the base and collector currents (I_B and I_C).

Calculate your transistor current gain (h_{FE}).

Calculate the d.c. current gain for two other very different values of base current.

Compare the three values you have obtained.

2 Set up the circuit on the next page.

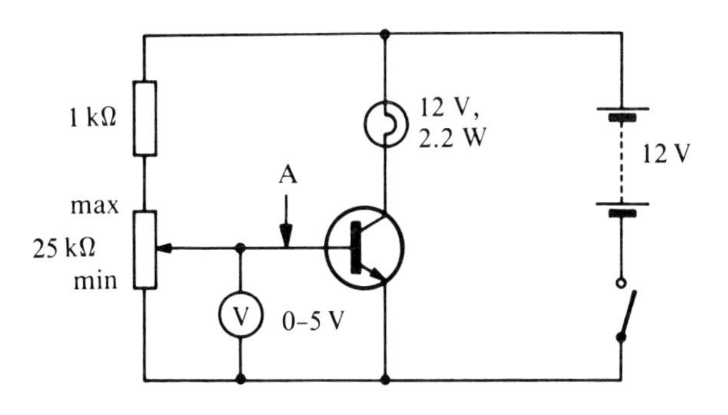

Copy the table below into your note book

Setting of potentiometer	*Base voltage* (volts)	*State of lamp*
Minimum		
10%		
25%		
50%		
75%		
Maximum		

3 Adjust the potentiometer to minimum and then switch on. Adjust the settings of the potentiometer to roughly those shown in the table and ***complete the table in your note book.***

4 Try adjusting the potentiometer between 0 and 10%. ***Note the significant results you obtain in this region.***
You may be able to think of ways to modify the circuit in order to make it easier to work in this confined region.

5 Fit an ammeter in the base circuit of the transistor at (A) and repeat the measurements in section 2 but ***note the ammeter readings.***

6 Replace the lamp with a relay and indicator bulb. Observe the effect of adjusting the potentiometer from minimum to maximum.
Draw this circuit in your note book.

Set up the following circuit:

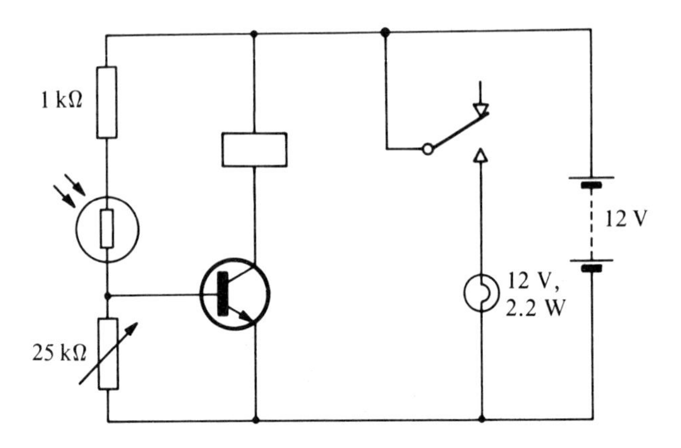

NOTE: The 1 kΩ fixed resistor must be fitted in series with the light dependent resistor in order to protect the photocell from excessive current flow when a light is shining on it. This fixed resistor will also limit the transistor base current and the normal 50 Ω base protection resistor may be omitted.

1 Position a light source so that it shines onto the photocell from a distance of about 30 cm.

2 Slowly adjust the rheostat from minimum to maximum resistance. ***Note what happens.***

3 Adjust the rheostat to a position where the relay just operates, then block off the light source. ***Note your results.***

Investigate the following (adjust the rheostat to give best results in each case).

4 Set up a simple photocell circuit using a relay, a photocell, and a voltage supply. Compare its overall performance with your transistor version.

Record your findings.

5 Reverse the positions of the 25 kilohm rheostat and the photocell in the transistor circuit.

   Compare the performance of this modified ciruit with the original transistor arrangement.

6 Replace the relay with an electric motor, and investigate the effect when the amount of light hitting the photocell is varied. ***Note what you find.***

7 ***In your notebook explain, in terms of transistor base and collector currents, how your two light-operated switch circuits function.***

Logic 1

1 A 'make' switch

is normally open (i.e. open when not operated). Set up the following circuit:

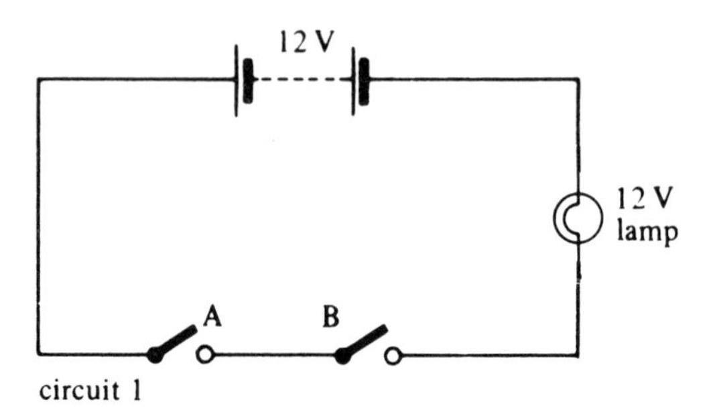

When the lever of switch A or B is 'toward' the sockets (not operated), call the operation 'O'. If the switch lever is away from the sockets (operated), call the operation '1'. If the lamp lights, call the result '1' (i.e. there is an output). If the lamp does not light, call the result 'O' (i.e. there is no output).

The lamp is illuminated for certain positions of the switches and remains unlit for other positions. ***Construct a three-column table for this circuit to include all possible positions (1 or 0) of the switches A and B. Indicate in the third column (Z column) the state of the lamp (1 or 0).***

A (input)	B (input)	Z (output)

The table you produce is called a 'Truth table'.

Suggest a name for this series arrangement of normally-open switches.

The name you choose should suggest which of the inputs (A, B) gives an output.

2 Set up this circuit and construct a Truth table as you did in 1. ***Suggest a name for this parallel arrangement of normally-open switches.***

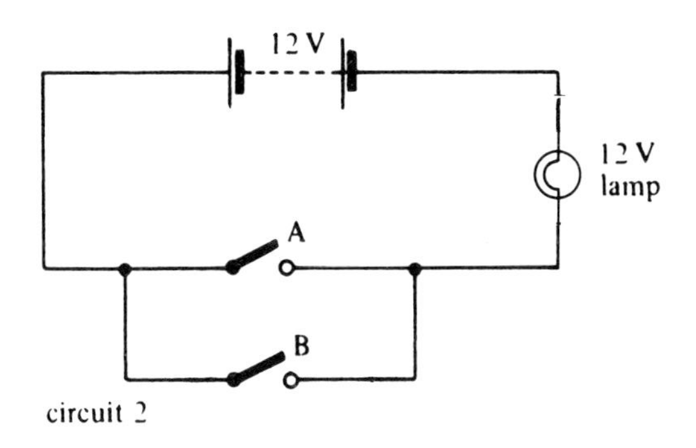

circuit 2

3 A 'break' switch is normally closed (i.e. closed when not operated). Set up the following two circuits and ***produce Truth tables.***

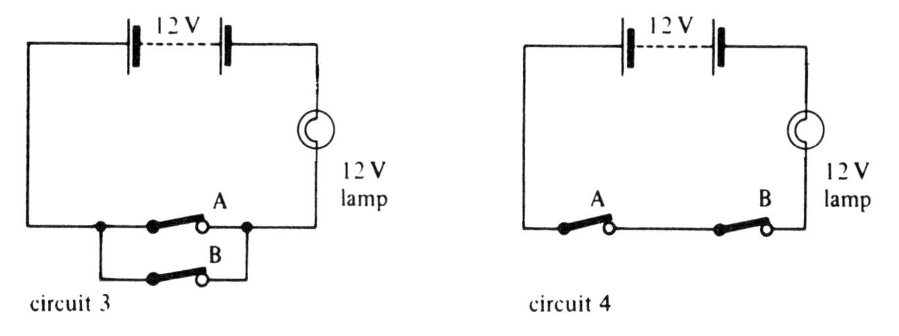

circuit 3 circuit 4

As before, if the switch is not operated, call the operation 'O', and if a switch is operated, call the operation '1'. In other words, the inputs are either '1' or '0'.

Compare the Truth tables of circuits 1 and 2 with those of circuits 3 and 4. ***Now suggest names for the arrangements 3 and 4.***

1 Set up the following circuit, ***draw a Truth table, and suggest a suitable name for the logical arrangement.*** As before, if a switch is unoperated, call the input 'O', and if it is operated, call the input '1'. An illuminated lamp represents a 'l' output, and an unlit lamp a 'O' output.

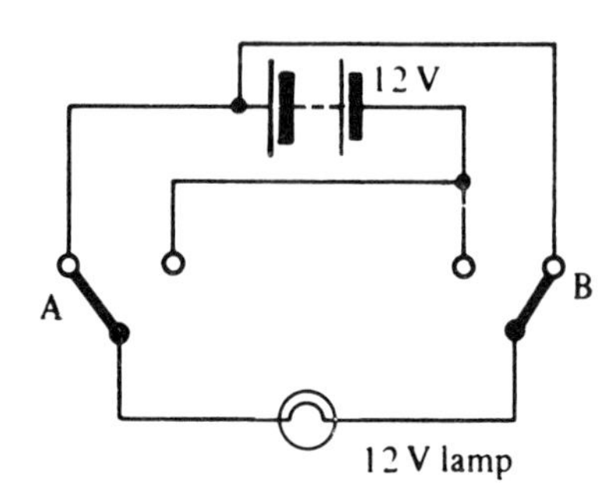

When your circuit is connected as shown, check that the switch levers are 'toward' the sockets.

2 Now set up this very similar circuit and ***draw its Truth table. Again suggest a suitable name.*** Again, when your circuit is connected check as shown, that the switch levers are 'towards' the sockets.

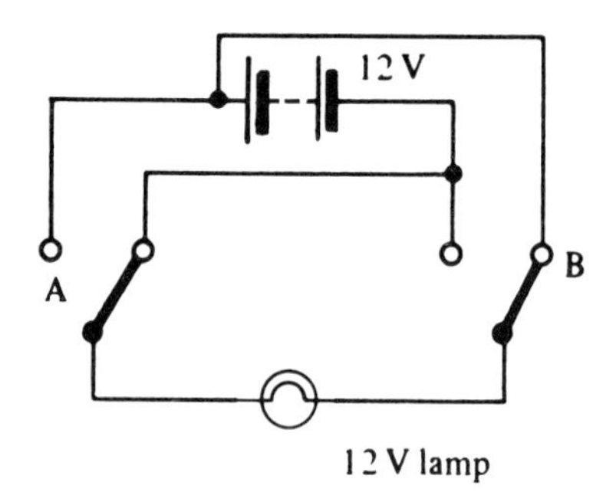

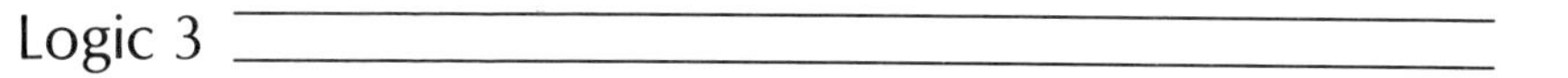

Logic 3

Important All the logic units used in this course require a 5 volt d.c. supply and will be destroyed by higher voltages. Your teacher will give you a suitable voltage supply unit — use this at all times with the logic units.

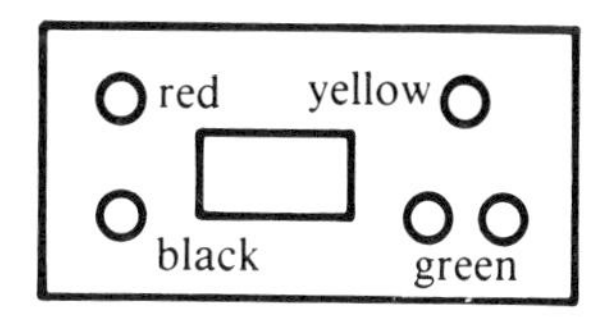

You have been given four logic units: AND, NAND, OR, and NOR.

Examine the AND unit. It should look like the diagram above.

Connect the 5 V d.c. supply to the red and black sockets (red to the positive terminal and black to 0 V). This must be strictly adhered to on all logic units where red and black sockets are found.

1 Connect the logic unit to a light emitting diode unit (LED), by connecting the *yellow* OUTPUT socket on the logic unit to the *green* socket of the LED. The *black* socket on the LED unit must be connected to 0 V. Connect the microswitches to each green input as shown, with the n/c contact to 0 V.

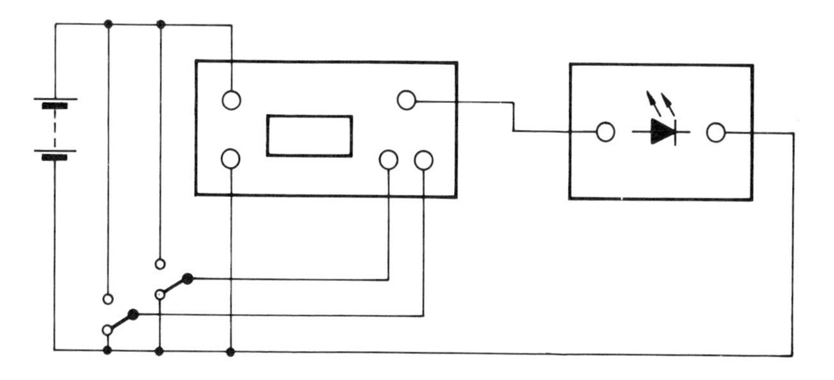

Does the LED glow?

2 Remember that with these logic units an input is '1' if it is connected to + 5 V and '0' if connected to 0 V.

Complete the Truth table for all possible input states.

A (input)	B (input)	LED state (output)

3 Repeat for the three other logic units.

In each case draw a Truth table to show what happens to the output for the various input states

4 Linking both inputs on the NAND unit together, connect them to 0 V.

Record what happens.

Connect the inputs to + 5 V.

Record what happens.

Can you think of a suitable name for a NAND unit connected in this way?

1 Using a transistor, resistors, diode units and a **12 volt supply** connect up the following circuit.

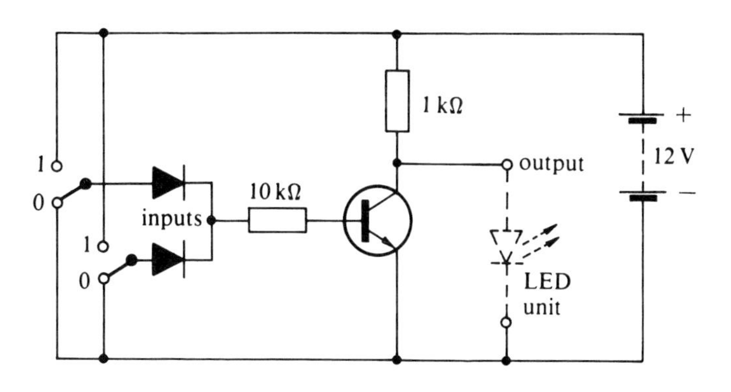

Construct a truth table for this circuit. What type of gate is it?

Add a second transistor to the previous circuit as shown below.

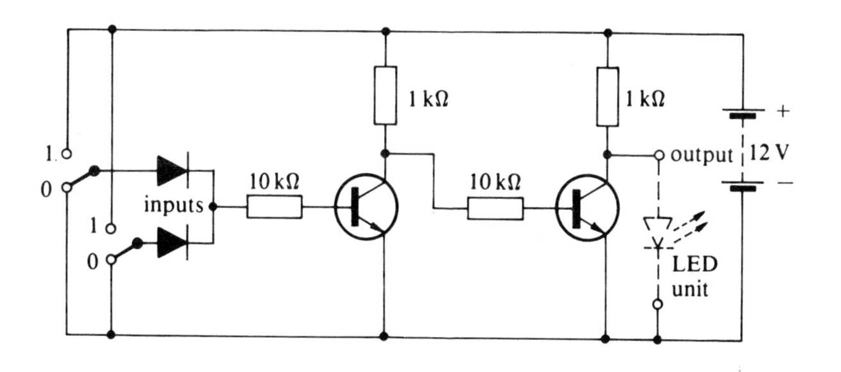

Construct a truth table for this circuit. What type of gate is it?

3 Connect up the following circuit. Note that the diodes are now reversed.

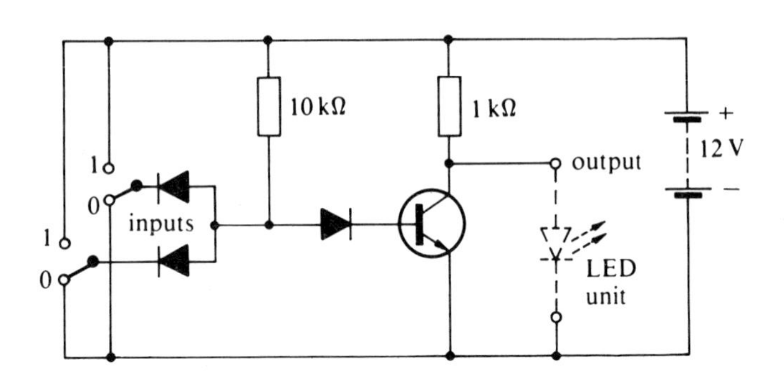

Construct a truth table for this circuit. What type of gate is it?

4 Add a second transistor to this circuit.

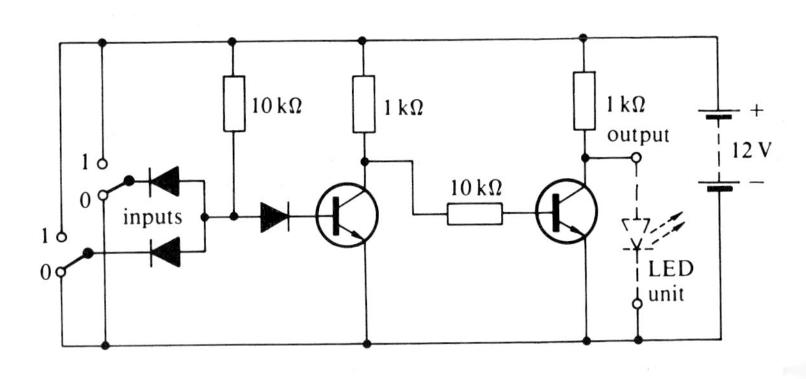

Construct a truth table for this circuit. What type of gate is it?

5 It does not matter whether a logic gate is made of toggle switches, circuits containing diodes, resistors and transistors, integrated circuits or pneumatic valves — the end result is the same. When we connect together several logic units to form a more complex arrangement we need to be able to show in a drawing what we have done. This is like drawing a circuit diagram in electronics. The five simplest types of gate are represented by the following symbols.

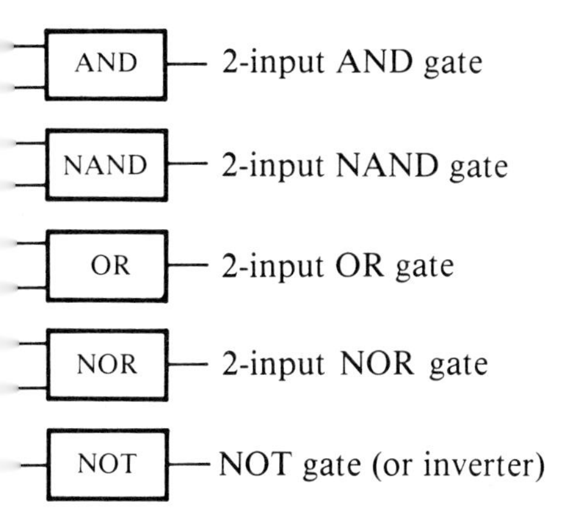

he symbols illustrated are of the 2-input type (except for the OT gate). However, it is possible to have more inputs. For xample:

opy the 2-input symbols and the NOT gate (with their names) to your book for future reference.

Logic 5

1 Using a logic units connect up the following logic arrangement (don't forget to use the 5 volt supply).

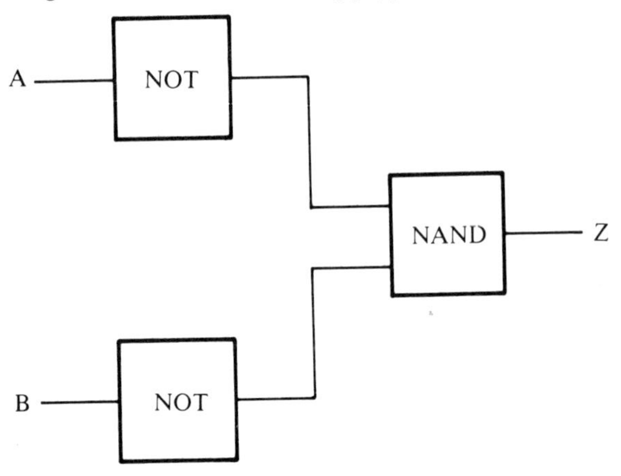

Copy the diagram.
Construct a Truth table for the arrangement.
What gate have you constructed?
Write a logic equation for the block diagram.

2 Connect up the following logic arrangement.

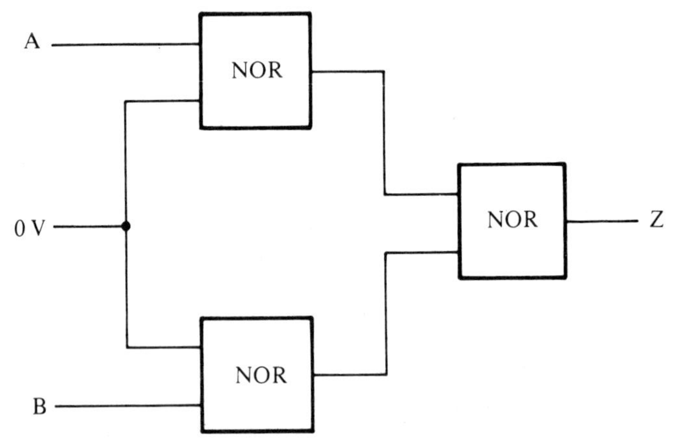

Copy the diagram.
Construct a Truth table for the arrangement.
What gate have you constructed?
Write a logic equation for the block diagram.

3 Connect up the following circuit.

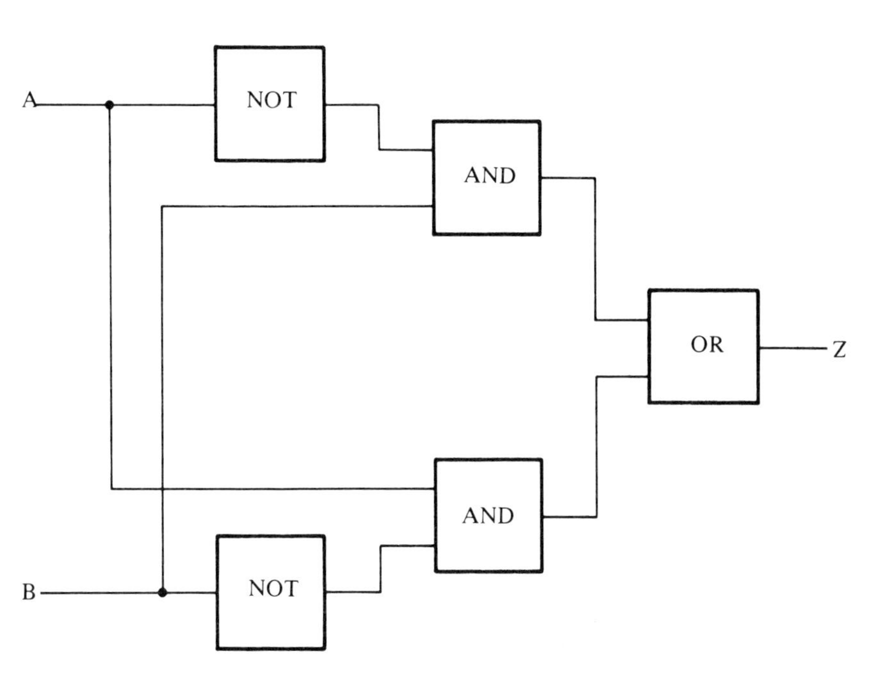

Copy the block diagram.
Construct a Truth table for the arrangement.
What gate have you now constructed?

4 A metal press in a factory has to have certain safety features built into it. It must operate only if manual switches A and B are pressed and the guard sensing switch C is operated. It must not operate if the light beam to a light-dependent resistor D is broken.

Write a logic equation for this system.

Construct a Truth table for the system.
Draw a block diagram of the logic arrangement.
Draw a possible electrical circuit diagram of the arrangement based on the use of retays.

Build and test your circuit.

The Approach To Project Work

The words in ***bold sloping*** type indicate the need for a written record of the work.

1 ***State*** the problem precisely.

2 ***Outline*** several possible solutions.

3 Choose one solution and ***say why*** you chose it in preference to any other.

4 Divide the problem into several separate sections. ***Work on each independently***, keeping in mind that all sections must fit together.

5 With the aid of ***sketches***, make a plan of how you intend solving each section of the project.

6 ***Suggest*** any modifications during or after construction.

7 Test the items which have been constructed. ***Say*** where your design has succeeded — and how well — and where it has failed.

8 ***Suggest*** further modifications for improvement. These may or may not be carried out.

The Approach to Investigational Experiments

1 The aim should be clearly stated.

2 In many investigations there are several factors or conditions which can vary. In a particular investigation, the variables should be limited to two — all other values which could vary must be kept constant throughout the investigation.

3 Draw a clear, labelled diagram, and briefly outline the method used.

4 *All* members of the group must record the results as soon as they are taken.

Whenever possible, recheck the results and, having done so, do not reject those which appear to be incorrect, since unexpected results are often obtained in experimental work.

If the results are tabulated, ensure that the quantities or values being measured ***and their units*** are indicated at the head of each column.

The conclusion arrived at is the most important part of the investigation. Study very carefully all observations and results or graphs, and draw as many conclusions as possible.
The conclusion should be a summary of what you have discovered.

The Use of Graphs

When a set of results is obtained from an experiment or investigation, it is often necessary to plot these on a graph to discover any relationship between two variables. On occasions you will find that there is no simple relationship, and that all one can say, for example, is that 'if the value of one variable is doubled, the value of the other is more (or less than) doubled'.

The following notes will help you to interpret graphs which show common relationships between two variable 'quantities'.

The straight line

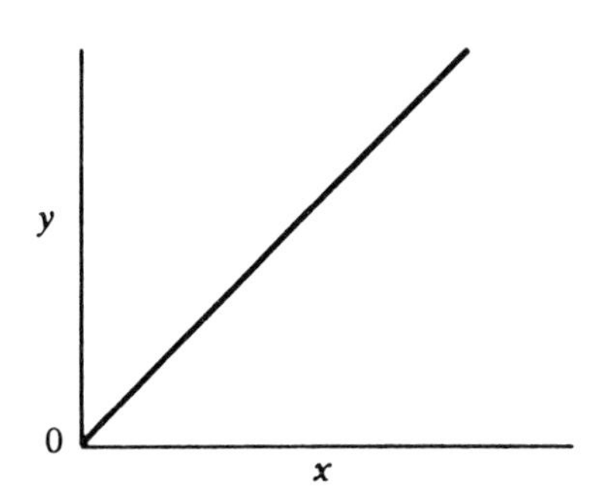

1 i) If the straight line passes through the origin (0,0) then x and y are in proportion, i.e. if x is doubled then y doubles.

ii) Often the slope (gradient) of a graph can provide useful information — the slope is obtained by dividing the difference between two values of y by the difference between the corresponding two values of x.

Graph to show the distance a car travels in a certain time

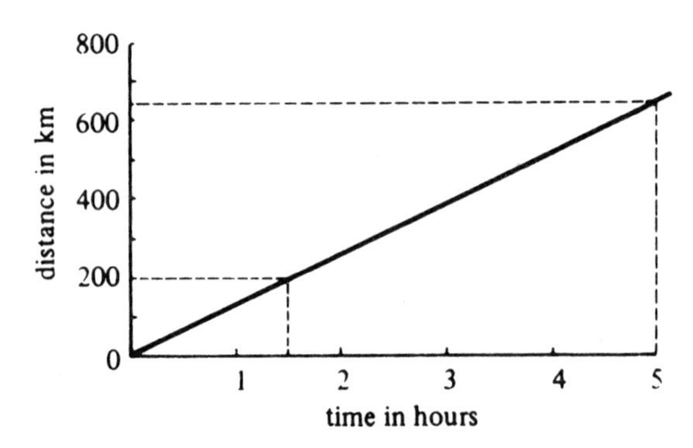

In this case the gradient is the speed of the car, i.e. $133\frac{1}{3}$ km/h.

NOTE: For accurate determination of a gradient, as much of the graph as possible should be used, so that any small errors you are likely to make in reading the values of x and y will be only a small percentage of the total readings.

iii) If a straight-line graph does not pass through the origin but makes an intercept on either of the axes, then some fixed quantity is always present in addition to values of x and y. For example, weights are hung from a spring, and you plot *the length of the spring* against the load attached to the free end.

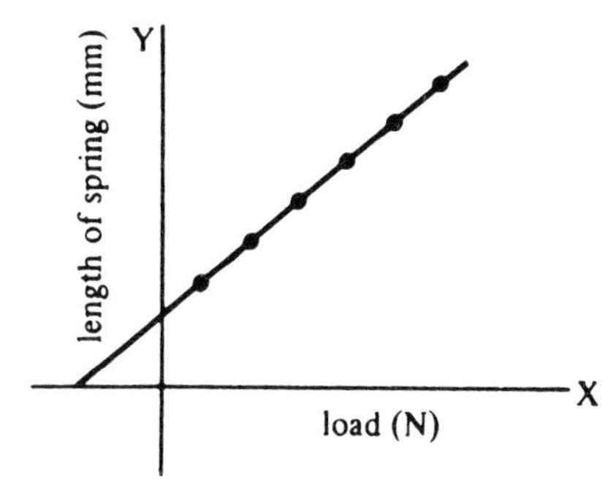

The intercept on the y axis (length of spring) gives the length of the spring with no weight added.

The intercept on the x axis (load (force) added) indicates the force theoretically needed to reduce the length of the spring to zero if conditions remain the same. (In this case the coils of the spring would touch first, and therefore this has no meaning). The slope of the graph gives the extension of the spring per newton of force.

2 When plotting any graph, it is normal practice to plot on the x axis the quantity you are varying, and on the y axis to plot the other quantity which is varying as a result.

3 Only rarely do all points appear exactly on the curve of a graph. Estimate whether the line is a straight line or a curve, and draw *the best line*. This means that there should be an equal number of points either side of the line and on average equally spaced from it. Your line then has the effect of *averaging out any unavoidable experimental errors*.

e.g.

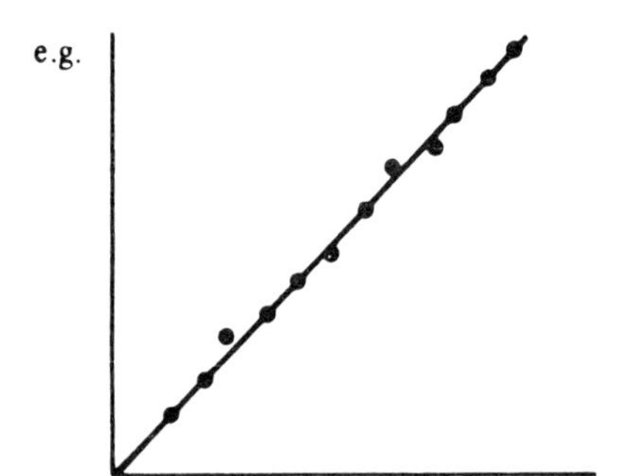

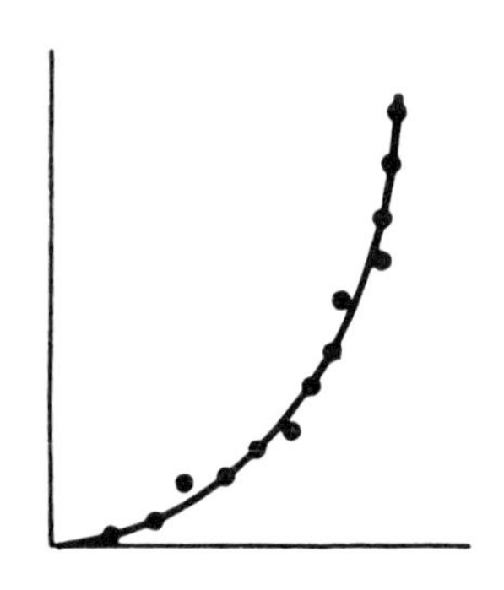

4 Choose scales which occupy as much of the graph paper as possible — a large graph is more accurately plotted and read than a small one.

Graphs showing proportion to the square of a quantity

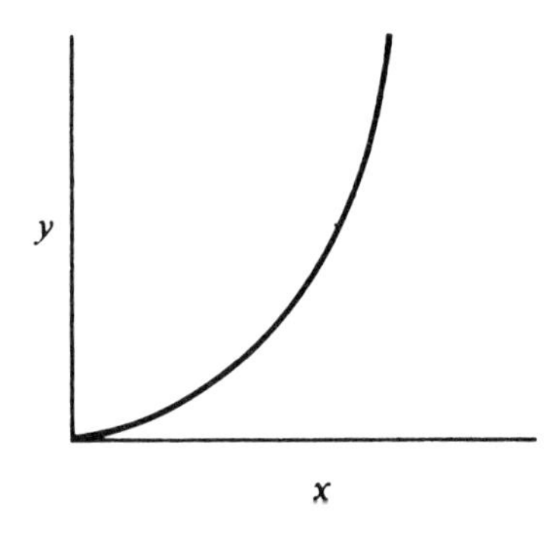

5 a) If you obtain a graph of the shape shown above, it may be a graph of $y = x^2$ (a parabola). Square the values you have plotted on the x axis and plot these against the same values you originally used for the y axis. If you get *a straight line*, passing through the origin, then the quantity you have plotted on the y axis *is proportional to the square of the quantity you have plotted on the* x *axis, i.e. if* the quantity plotted on the x axis is *doubled*, then the quantity plotted on the y axis is *quadrupled.*

Alternatively, read off a value for the quantity you have plotted on the x axis and find out if by doubling it you quadruple the quantity you are measuring on the y axis. (Repeat this for different values of x.)

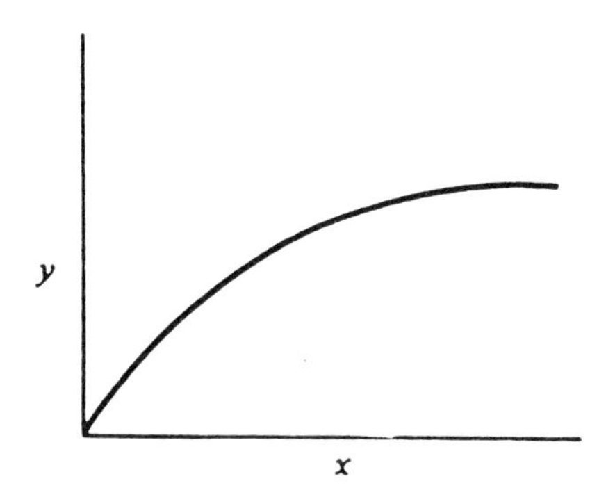

b) If you obtain a graph of the shape shown above, it may be a graph of $x = y^2$ (a parabola). Square the values you have plotted on the y axis and plot these against the same values you originally used for the x axis. If you get a *straight line*, passing through the origin, then the quantity you have plotted on the x axis *is proportional to the square of the quantity you have plotted on the y axis*, i.e. if the quantity you have plotted on the y axis is *doubled*, then the quantity plotted on the x axis is *quadrupled.*

Alternatively, read off a value for the quantity you have plotted on the y axis and find out if by doubling it you quadruple the quantity you are measuring on the x axis. (Repeat this for different values of y.)

Graphs showing inverse proportion

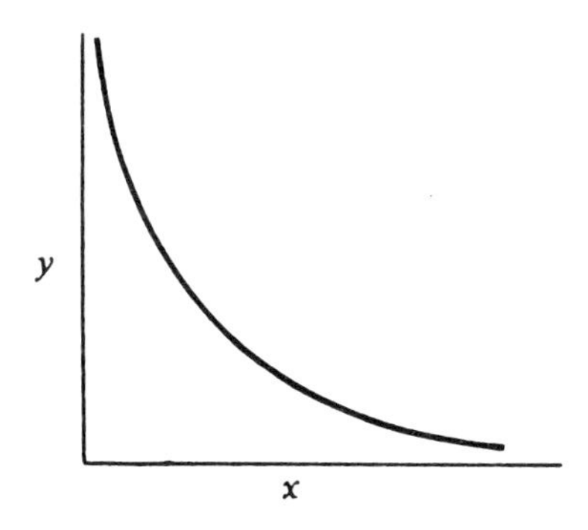

If you obtain a curve of the shape shown above, it may be a graph of $y = 1/x$ (or $x = 1/y$); i.e. if you double y you halve x, and vice versa. (A graph of $y = 1/x$ is called a 'rectangular hyperbola'.) Find the reciprocals of the values you have plotted on the x axis, and plot those values against the values you originally plotted on the y axis. If you obtain a straight-line graph passing through the origin, then *either one* of the values you have plotted *is inversely proportional to the other*.